THE SCREENING AND GRADING OF MATERIALS

BY

J.E. LISTER

CHEMICAL ENGINEERING SERIES

Wexford College Press

2007

INTRODUCTION

THE Screening and Grading of Materials form a very important part of many modern manufacturing operations, and are essential in order to separate the manufactured or raw material into various grades according to their suitability for different purposes, and to remove any impurities present as far as possible in order to enhance the value of the product.

Coal, coke, ores and sands, for instance, must be carefully graded and cleaned, so that the demands of various industries may be met satisfactorily, and the more thoroughly and carefully this operation is carried out, within practical limits, the better will be the demand for the various grades of the material. The removal of impurities should, as far as possible, be carried out at the same time as the grading or classification into sizes, in order to reduce the amount of handling required, as in most cases the cost of the operation of grading and cleaning will vary practically in proportion to the amount of handling to which the material is subjected. Many types of machinery have been devised to facilitate these important operations, and it is proposed to describe, in the following pages, the more important types in general use in various industries, with some account of the class of material for which they are most suitable.

In many cases the screening and grading will form merely one step in a series of operations by means of which the material is conveyed from the point of production to the point of dispatch, but it is not intended in the present work to deal with the problem of the transportation of the material where this does not form an integral part of the grading operations.

In some cases screening and grading machinery is designed specially for dealing with a particular class of material, but in many other instances one type of apparatus is suitable for various purposes, and it will, therefore, be most convenient to describe the different types in detail and to give in connection with each some account of its use in practice.

The amount of grading that can economically be carried out in any instance will depend on the value of the product and the cost of the operation, and it is therefore important, when deciding on a scheme for grading or screening any material, to consider the relative advantages and cost of up-keep of the different types of apparatus that may be suitable, and it is hoped that the descriptions given in the following pages will be of assistance in enabling a selection to be made.

As an example of the importance of proper screening and grading of material, the case of coal may be mentioned. Coal forms the basis of many of our most important industries, and the various sizes and qualities have each their own particular use.

The purchase of coal by large consumers is now being carried out more generally on the basis of its calorific value, and although this depends largely on the source from which it is obtained, it is also affected to a great extent by the amount of incombustible matter present, and steps are therefore taken to remove as much as possible of the shale and other incombustible material while the coal is being graded into its various sizes.

The introduction of inexpensive methods by means of which the coal may be purified and classified is of great importance, as the installation of plants employing such methods will enable a much greater quantity of coal to be mined from a given area economically, whereas without such methods a large quantity of combustible matter cannot be raised owing to the amount of foreign incombustible matter intimately mixed with it.

Coal-washing plants are now generally installed as part of a modern colliery equipment, but pneumatic separation plants are being developed and promise to be very successful for certain classes of work.

The grading of the coal takes place at the pit-head, the tubs of coal drawn from the pit being discharged over screens in order to separate· it into various sizes in readiness for the subsequent operations.

The larger sizes of coal are delivered on to picking belts or tables where stones, shale and other foreign materials are removed by hand, while the smaller coal, which has passed through the preliminary screens, is conveyed to the washery, where the " dirt " is removed by means of water.

Coal above $1\frac{1}{2}$ inch or $1\frac{3}{4}$ inch cube is generally hand-picked, and sizes below this are treated in the washery. The arrangements actually adopted to effect the separation of the various sizes and the elimination of the dirt will be described more fully later.

Various classes of sands and gravel are very largely used in modern industry, and among the many materials of which they form an important part may be mentioned bricks, cement and concrete, glass and pottery, plastic, insulating and refractory linings for furnaces, while sand is also used extensively for filters, sand blasting and the formation of moulds in foundries.

For all these uses it is important that the sand, of whatever kind, should be properly graded, and various types of machinery are used for the purpose.

Many sands are contaminated by clay, while others are affected by the presence of iron or other compounds which would impair the quality of the finished product.

In most cases mechanical treatment is sufficient to effect the necessary grading, but for some purposes the sands must be burnt or treated chemically in order to get rid of the impurities. Treatment with water will remove clay or silt, but is not sufficient for removing iron contamination, and it is generally necessary, after a material has been washed, to remove the surplus water before the sand can be used.

Many mineral-bearing sands and fine ores have to be concentrated by water before their final treatment for the extraction of the mineral. This process comprises the washing of the ore in suitable vessels in such a way that the lighter siliceous material is removed while the more valuable and heavier portions of the ore are retained.

Mechanical screening or grading into sizes may be employed satisfactorily for all materials exceeding $\frac{1}{400}$ inch diameter. For dealing with material in an extremely fine state of division air separators are now being used successfully, while electrical methods are also being utilised for dealing with magnetic substances and for removing dust from blast furnace and similar gases to render them fit for combustion and to enable the valuable constituents of the dust to be recovered.

In the manufacture of Reinforced Concrete, which now plays a very important part in the industrial world, it is becoming recognised that the proper grading of the sand used is vital for ensuring successful construction. It is not only essential that the sand be clean and free from clay, but there must be the correct proportion of the various sized grains in order that all the voids may be filled, and to ensure that the completed structure may form a solid homogeneous mass.

J. E. L.

1*

CONTENTS

INTRODUCTORY

xi

LIST OF ILLUSTRATIONS

CHAPTER 1

INCLINED SCREENS

The simplest form of screen in common use consists of a frame on which is fixed a piece of wire mesh, perforated plate, or expanded metal. The frame is set up at an angle and the material to be screened thrown against it by hand. A screen of this kind is only used where comparitvely small quantities of material are to be handled, and is only suitable for dealing with material of moderate size and for seperating it into two grades.

This type of screen may be made suitable for dealing with larger quantities of material by feeding it mechanically over the top of the screen by means of a conveyor or elevator in any convenient manner, the smaller particles passing throgh the meshes and being collected by a delivery plate or chute, while the larger pieces pass over the openings and are deposited at the foot of the screen.

The material should not fall directly on to the screen, but on to a solid "dead plate" from which it slides on to the screen plate. This arrangement assists in distributing the material evenly over the screen and reduces the amount of wear and tear.

Plain inclined screens of this type are not very efficient unless the material is fed on very thinly, as otherwise it may slide over the screen in a solid mass, carrying some of the fine particles over the openings. In any case, the size of the pieces passing through the screen will be somewhat less than the actual size of the openings in the perforated plate or wire mesh.

In some cases hinged flaps have been arranged over the top of the screen. The weight of these flaps slightly retards the flow of the material and prevents it from passing over the screen openings at excessive speed, and at the same time helps to spread the material evenly.

The slove of the screen must be steeper than the angle of the repose of the material to be handled, taking into account the resistance caused by the type of screen plate or mesh which is used, otherwise the material will hange up and the continuous action of the screen will be interrupted.

If the material is likely to vary in nature or moisture content, provision should be made for altering the inclination of the screen from time to time as the material varies.

When very fine material is to be separated it is advisable to have some means of agitating the screen in order to prevent clogging of the openings.

Where agitation is employed the inclination of the screen may be less than in the case of a fixed screen handling the same material, as the passage of the particles over the screen is assisted by the movement.

For materials which are free from long flat

pieces, straight bars or wires may be used, either round or rectangular. For moderately large material perforated plates are best, though the effective area is less than when bars are used, and this fact should be taken into account when designing the dimensions of the plant. The plates must be of adequate thickness in order to be sufficiently strong to bear the weight of the material, after making allowance for wear and for the metal removed in making the perforations.

The nature of the material to be handled must also be taken into consideration, as some materials, such as coke, are extremely abrasive and will wear away the metal rapidly.

It is an advantage to have the holes made slightly tapered, so that any pieces which have passed through the opening can immediately fall away freely.

Precautions should be taken that the material is spread out evenly over the full width of the screen, which must be of ample length, otherwise some of the finer material will be unavoidably carried over and discharged with the tailings.

A bar screen offers less resistance to the flow of the material than a perforated or wire mesh screen, and may therefore be arranged at a flatter angle, but the disturbance of the layers of material passing over it is not so great as with the perforated screen and the separation of the various sizes is therefore not so efficient.

An inclined screen of this type may be made up of two or more plates having different sized openings the smaller being at the top, so that the material may be separated into several different sizes.

For screening flour and other fine material, bolting silk, stretched on a suitable frame, is used, or very fine woven bronze wire.

In the design of an inclined Bar Screen there are a number of points to be observed if the best results are to be obtained. It is essential that the bars should be of ample length in order to give the material time to separate out, and they should be made of taper section so that any particles passing through the openings are free to fall away without any tendency to jam, while it is also advisable that the bars should be slightly wider at the upper end of the screen than at the bottom. The increasing width of the openings offered to the material resulting from this arrangement makes the screen self-cleaning, as any pieces which may jam in the slots are released by the pressure of the following material, which forces them into a wider part of the slot. This self-releasing property is of particular value in dealing with material which may contain long tapering pieces, as in the case of coke.

The upper ends of the bars should be flush with, or covered by, the dead plate or the bottom plate of the shoot in which the screen is incorporated, to prevent any tendency for the material to lodge against the bars, and the bottom ends should be turned over to an amount not less than the depth of the bar, so that there is no obstruction at this point.

The design shown below (Fig. 1) has been found

very successful in practice, particularly for dealing with coke.

Fixed bar screens are generally used for the preliminary screening of coal as it is raised from the pit, though screens of the jigging type are frequently used and in some cases a combination of the two types is installed, the coal from the fixed screens passing directly to shaking screens

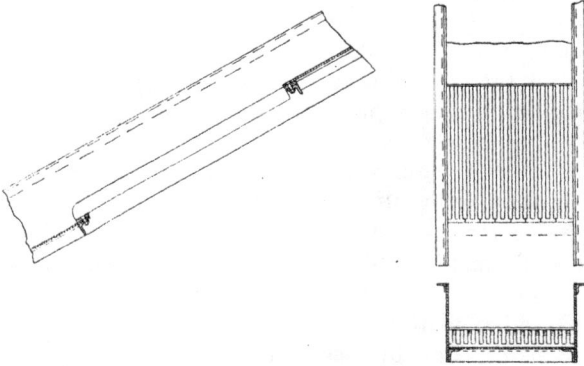

Fig. 1.—Arrangement of Inclined Bar Screen.

in order to effect a more complete separation of the various sizes.

The angle at which the bar screens are arranged depends on the size of the coal which it is desired to retain on the screen, as the smaller the coal the greater must be the inclination for satisfactory operation.

The following table gives the angles which have been found satisfactory for various classes of coal :—

Run of mine coal (*i.e.*, as delivered from the colliery
    tubs) .   .   .   .   26°
Cobbles .   .   .   .   23–24°
Nuts   .   .   .   .   24–25°
Slack or dust .   .   .   .   .   .   32°
Cannel .   .   .   .   23–24°

Shoots should be arranged at the lower ends of the screens to obviate breakage as far as possible and thus to prevent small coal produced by breakage from being mixed with the larger grades after they have been classified. The angle of the shoot should vary from 30° to 45°, according to the size of the coal, the steeper angle being used for small coal, which has a smaller angle of repose than the larger lumps.

For other materials the angle of inclined screens may vary from 30° to 60°, the lower figure being suitable for washed gravel, coarse dry sand and cement clinker, while the higher is suitable for fine cement, moist sugar, salt, small washed coal and similar materials.

Rocking Bar or Oscillating Screens are particularly suitable for dealing with material of fairly large size, and are generally used in connection with crushing plants for coal or rock for the purpose of separating out the smaller pieces to avoid putting them through the crushers. This arrangement increases considerably the capacity of the plant, as the crusher is not called on to deal with a mass of small material which is already below the size to which it is intended to reduce the larger lumps. In this type of screen two sets of bars are arranged at a slight incline, usually 15° to 20°, the bars of one set alternating with those of the other. The lower ends of the bars slide in guides, while the upper ends of each set are connected to a separate crank or eccentric so arranged that the two sets of bars move forward and backward, one set moving forward while the other set is returning. The bars are provided with notches or slots of a size suitable for passing the material which is already smaller than the maximum size of the product. The reciprocating motion of the bars, combined with the angle at which they are set, causes the mass of material to move forward down the incline, thus ensuring a regular feed to the crusher, while the small material passes through the openings into a separate shoot which leads it back into the product from the crusher.

In a modified form of this type of screen the two sets of bars have lugs cast on either side of a central rib, the lugs on one set of bars coming midway between those of the other set, thus forming a series of rectangular openings. The bars are all pivoted on a common axis or shaft at the lower end of the screen, while at the upper end the two sets of bars are connected to the two horizontal arms of a T-shaped lever. The vertical arm of this lever is connected by a rod to a crank or eccentric, and as the shaft carrying the eccentric revolves the two sets of bars move up and down alternately.

This type of screen is very effective, as the up-and-down motion of the two sets of bars agitates the material and causes thorough separation of the smaller pieces, while the larger lumps are fed forward over the lower end.

These screens are often combined with a reciprocating feeder which delivers a regular supply of material over the bars. The arrangement of this type of screen is shown in Fig. 2. The width is usually 3 to 5 ft. and the speed 100 to 150 r.p.m. The stroke of the reciprocating type may be from 5 to 9 inches, the power absorbed $1\frac{1}{2}$ to 3 B.H.P.

When fine materials have to be dealt with it has been found essential to adopt some means for

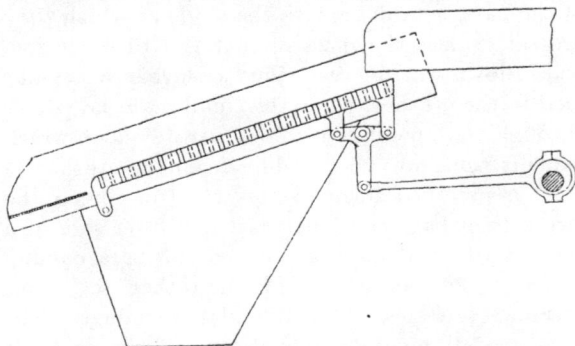

FIG. 2.—Arrangement of Rocking Bar Screen.

rapidly agitating or vibrating the screens, otherwise the material has a tendency to adhere to the screening cloth or wire and clog up the openings.

Inclined screens of this type are largely used for the screening of cement, superphosphate of lime and similar materials which must be reduced to a very fine state before use.

The vibration of the screens is obtained in various ways, generally by tapping the screen at rapid intervals by suitable mechanical means or by reciprocating the frames by means of an eccentric, though in some cases electrical methods are employed.

The screens or sieves are arranged at an angle of about 45° and are enclosed in a casing of wood or steel plate, along the top of which runs a worm conveyor adjusted to give an even feed over the whole width of the screen.

The fineness of the screened product depends mainly on the size of the screen openings, but the angle of inclination of the screen has also some influence on the size of the material passing through, the steeper the angle the smaller the product for a given size of mesh. In any case the size of the product is considerably less than the actual size of the openings.

The Newago Screen, made in this country by the Sturtevant Engineering Co., Ltd., of London, is a typical example of the mechanically vibrated screen operated by a tapping mechanism. It consists of a steel casing which is practically dust-proof, arranged at a suitable angle, generally about 45°, and containing one or more screens, mounted on steel frames which are supported on springs. The screen cloth is also held in position by springs which maintain a strong tension on the cloth and keep it uniformly taut in the frame. This is an important point in the design of the screen, as it adds materially to the sharpness of the vibration, which is essential for obtaining efficient separation of the material. The front of the steel casing is hinged to facilitate the insertion and removal of the screen frames, and inspection doors are provided for examining the material when necessary. Large screens are divided into sections.

The material to be screened is fed into one end of a rectangular box extending along the whole of the top of the screen. This box contains a spiral conveyor with an adjustable spill board or weir for ensuring a regular distribution of the material over the whole width of the screen.

Across the front of the casing are arranged a number of shafts, each carrying a series of cranks or cams which lift at rapid intervals a corresponding number of hammers. As the shafts, which are driven by belts from an extension of the conveyor spindle, revolve, the cranks move away from the hammers, which are allowed to fall freely on to the upper ends of pins which are connected at their lower ends to the screens. The screens are therefore struck at rapid intervals, causing vibrations to be set up in the taut screen wires. These vibrations are of small amplitude, to prevent the material jumping off the screen, but they occur very rapidly.

When two or more screens are used, one above the other, to separate the material into several grades, the pins are carried through and attached to all the screens, so that they all vibrate in unison.

The rate of feed is adjustable by means of the longitudinal board in the conveyor casing.

The power required for operating a screen of this type is small, the largest screen only requiring about 1 B.H.P.

The diagram (Fig. 3) shows the general arrangement of a double screen Newago Separator.

These screens are suitable for materials varying in size from $1\frac{1}{4}$ in. to 180 mesh. As mentioned

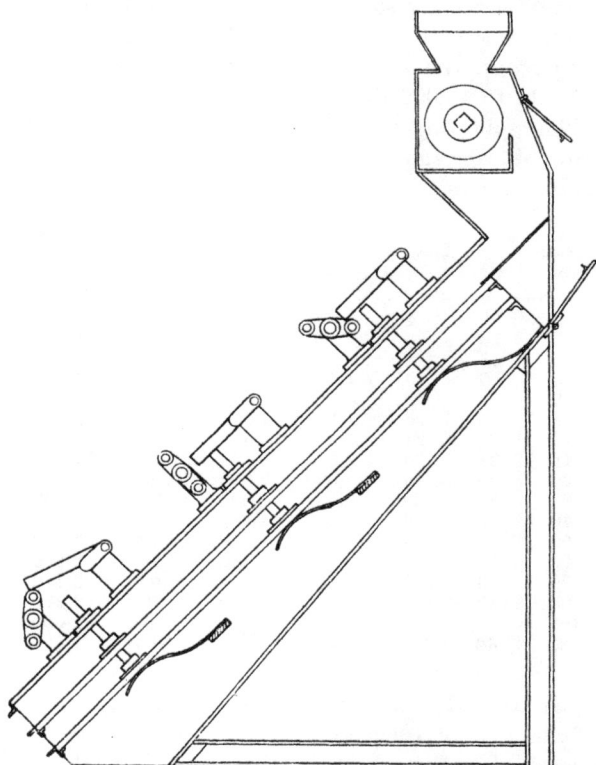

Fɪɢ. 3.—Sturtevant-Newago Screen Separator.

above, the actual size of the particles passing through the screen is considerably less than the size of the openings, a 10-mesh screen producing approximately a 20-mesh product, while a

120 screen gives approximately a 200-mesh output.
The screening surface varies from 4 ft. × 6 ft. to 12 ft. × 6 ft., and the capacity, which depends largely on the size of the material, varies from 450 lb. to 46,000 lb. per hour.

The following table gives the approximate output of Newago Screens of various sizes with different meshes and gauge of wire :—

Mesh of cloth.	Size of wire.	Approx. mesh of output.	Screen 4'×6'.	Approx. screen 6'×6'.	Capacity screen 8'×6'	Lbs. per hour screen 12'×6'.
2	11	4	11,650	23,300	35,000	46,600
3	13	6	11,000	22,000	33,000	44,000
4	14	8	10,300	20,600	31,000	41,200
5	15	10	9,650	19,300	29,000	38,600
6	16	12	9,500	19,000	28,000	38,000
8	18	16	8,160	16,320	24,500	32,640
12	21	24	7,330	14,660	22,000	29,320
16	24	30	6,650	13,300	20,000	26,600
20	26	40	6,000	12,000	18,000	24,000
30	31	60	3,650	7,300	11,000	14,600
40	33	80	2,300	4,600	7,000	9,200
60	36	100	900	1,800	2,700	3,600
70	37	120	700	1,400	2,200	2,800
80	38	140	600	1,200	1,900	2,400
90	39	160	550	1,100	1,700	2,200
100	40	180	500	1,000	1,500	2,000
120	42	200	450	900	1,400	1,800

In the Hummer Screen made by the W. S. Tyler Co., of Cleveland, U.S.A., the vibration of the screens is obtained by means of electro-magnets mounted in the casing over the screen frames. The electric circuit in which the magnet coils are connected is made and broken with great rapidity, causing the magnets to attract and release alternately blocks of steel attached to the screen. This action keeps the screen in a state of rapid vibration without an excessive movement which would cause the material to jump off the screen surface.

The whole arrangement is, as usual, enclosed in a dust-tight casing.

A modified form of Inclined Vibrating Screen has recently been introduced by the Sturtevant Engineering Co., Ltd., and is known as the Moto-Vibro Screen. The general construction is similar to that of the Newago Screen described above, but the vibration is obtained in a different manner, which gives to the Moto-Vibro Screen several advantages over the earlier patterns.

The screen cloth is attached to rigid steel frames, one, two or three of which may be arranged in the main casing to give various grades of material. These frames are vibrated at the rate of about 1800 double strokes per minute by means of an eccentric having a very small throw and driven by belt from a motor or line shaft.

The vibrating mechanism is situated underneath the dust-tight casing containing the screens, thus leaving the upper part of the box quite free for access to the screens. The vibration is transmitted directly to the frames and not to the screen cloth itself, thus avoiding excessive flexure of the wires of which the cloth is constructed. This arrangement has the advantage of vibrating the whole area of the screen uniformly, without leaving dead areas subject to little or no vibration, as in the case where the blows are transmitted directly to the cloth.

The position of the vibrating mechanism is so arranged that the maximum amount of vibration takes place at the upper end of the screen, where there is the greatest thickness of material. The fine material is thereby quickly separated from the coarser particles and the greater part of it passes through the screen in the upper zone. The remaining portion of the screen surface has then only to deal with a reduced quantity of material and is thereby rendered more efficient for separating out those particles which approximate in size to the openings of the mesh.

The amplitude of the vibration becoming less as the material passes down the screen, the particles are not projected so far from the surface by the vibrating action, and this again increases the efficiency of the screen in giving complete separation of the under- and over-size material.

The screen may be fed by means of a worm conveyor as described for the Newago Screen, or by means of a spout or shoot.

The power required for operating the Moto-Vibro is approximately 1 B.H.P. for all single section screens, and for two unit screens with spout feed, while 2 B.H.P. is sufficient for two unit screens with worm feed and for all three unit screens.

Each unit is approximately 6 ft. × 3 ft. in area, and the standard angle at which the screens are arranged is 35°, though angles from 30° to 40° may be used if necessary. Too steep an angle will cause the material to pass too rapidly down the screen, with the result that the fines are not completely separated out.

A similar type of machine designed for heavier duty is supplied with a screening surface of 4 ft. × 5 ft. and is generally used for handling comparatively large material down to 10 mesh. Such a screen will deal with 50 tons of damp sand per hour passing 25 to 30 tons up to $\frac{3}{16}$ in. in size through the screens.

This Moto-Vibro Screen is shown in Fig. 4.

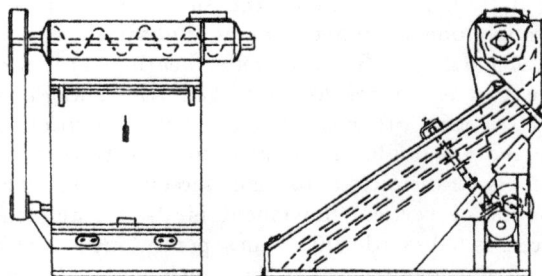

Fig. 4.—Sturtevant Moto-Vibro Screen with Conveyor and Feed.

The Hoyle Patent Vibrating Screen, manufactured by the Grange Iron Co., Ltd., Durham, has been designed specially for dealing with fine products, and is used for screening small coal, coke, breeze, slag, ores, sands, salt and other similar materials. It is made in two sizes, 6 ft. × 4 ft. and 4 ft. × 4 ft., the larger screen having a capacity of 50 tons and the smaller 35 tons per hour when fitted with a screen cloth having a $\frac{1}{4}$ in. square mesh.

The Hoyle Screen consists of a steel frame

mounted on trunnions in such a way that it can be adjusted to any required angle and clamped in that position. The screen cloth is clamped to the frame and held in tension by adjusting bolts. Steel bars are attached to the screen cloth and connected through adjusting blocks to the vibratory arms, which are mounted on eccentrics turned on the driving shaft. The driving shaft is carried in bearings secured to a crossbar forming part of the main frame, one end of the shaft being carried through one of the trunnions and fitted with a driving pulley. Ball and roller bearings are used. The two eccentrics are arranged at an angle of 180° to each other so that at each revolution of the shaft a double movement in a vertical direction is transmitted to the screen cloth. The amount of vertical movement is $\frac{1}{16}$ in., and the screen is vibrated 3000 times per minute. This rapid movement ensures the continuous agitation of the material passing over the screen and prevents any choking of the apertures. The separation is very efficient, and allows of wet or dry material being screened satisfactorily. The power absorbed is 2 B.H.P., and the output large for the area of the screen.

The Hoyle Screen is made with either a single or double deck, the arrangement of a single deck screen being shown in Fig. 5.

The Overstrom Vibrating Screen, supplied by Messrs. Hugh Wood & Co., Ltd., of Newcastle-on-Tyne, employs an entirely different principle for obtaining the vibration required for ensuring rapid screening and the prevention of the clogging

of the screen meshes. The vibratory driving
mechanism consists of an unbalanced pulley which
is secured rigidly to the screen frame, and revolves
at a speed of about 2000 r.p.m. for large screens,
up to 3000 r.p.m. for the smallest size.

The screen is supported by two hollow shafts,
the ends of which are carried beyond the sides of

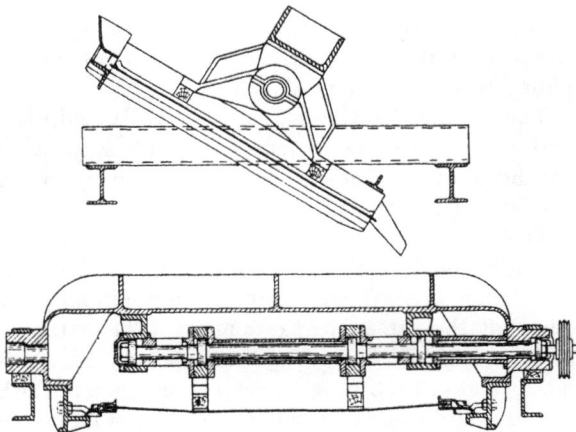

FIG. 5.—Hoyle Vibratory Screen.

the screen frame and supported by four spiral
springs carried in castings bolted to rigid supports. These springs allow the screen to be
vibrated without any of the vibration being
transmitted to the supporting structure.

The unbalanced pulley is carried on a third
stationary hollow shaft bolted to the screen frame,
and is provided with counter-weights which can
be changed as required for varying the amplitude

of the vibratory movement, the heavier the
weights that are fixed in the pulley the larger
being the movement of the screen. The amplitude of the movement may be varied from $\frac{1}{32}$ in.
to $\frac{3}{16}$ in. by this means.

The unbalanced pulley is lubricated by oil
under pressure, the pressure being produced by
centrifugal action caused by the rotation of the
pulley, and this arrangement ensures effective
lubrication at any speed, so that the wear on the
pulley bush is extremely small.

The motion of the screen is slightly elliptical
and the vertical movement is slightly greater at
the head of the screen, due to the pulley being
placed nearer this end.

The screens are attached to the box or frame,
which is constructed of wood, by means of angles
and bolts, liners being provided for clipping the
edges of the screen to the angles. This arrangement allows of the screens being stretched abso-
lutely tight, so that no part of·the screen cloth
can sag, and the whole screening surface moves in
unison. There are no projections on the screen
to cause any obstruction to the travel of the
material, and the whole surface of the screen is
therefore effective.

The screen frame can be arranged to carry one,
two or three screens, so that the material can, if
required, be separated into four different grades.
The various screens are inserted or removed quite
independently of one another.

The direction of the travel of the material
depends on the direction of rotation of the pulley,

which should be rotated so that the top of the pulley moves in the opposite direction to the travel.

The Overstrom Screen has been used successfully for handling coal, coke, rock, sand, gravel, marble dust, gypsum, salt, graphite and other materials, and can be used on either wet or dry products. It is made in several sizes, from 1 ft. 6 in. × 6 ft. to 3 ft. × 8 ft. and 4 ft. × 10 ft., and the driving power required varies from ½ H.P. for the small size to 2 H.P. for the 3 ft. × 8 ft. screen. The capacity varies from 1 to 1200 tons per 24 hours, according to the size and arrangement of the screen, the nature of the material handled and the size of the screen mesh.

The usual angle at which the screens are arranged for all purposes is about 37°. They will handle materials varying in size from 2 in. diameter to 200 mesh, or finer.

The following table gives the approximate output of the various sizes for different mesh screens in tons per hour :—

Mesh.	Screen 1' 6" by 6' 0".	Screen 2' 0" by 8' 0".	Screen 3' 0" by 8' 0".
2	16·875	34·375	50·00
3	13·75	28·125	41·25
4	10·625	21·875	31·25
6	9·75	18·75	27·5
10	8·125	16·25	23·75
16	5·0	10·625	15·625
20	3·73	8·125	11·875
30	2·5	5·625	7·5
40	1·875	4·375	5·625
60	1·5	3·125	4·375
80	0·875	1·875	2·625
100	0·625	1·5	2·125

The arrangement of a typical Overstrom Screen, with wooden framework, is shown in Fig. 6.

A Shaking Inclined Screen made by Messrs. Coal and Ore Dressing Appliances, Ltd., consists of an inclined trough, having a perforated bottom plate, carried at the delivery end on a pair of rollers, and supported at the head end by means

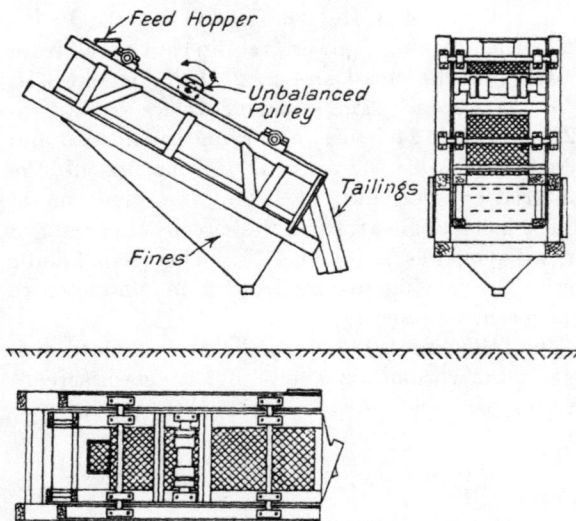

FIG. 6.—Overstrom Vibrating Screen.

of links from a fixed structure. A steel tee is fixed underneath the trough, and this is struck at rapid intervals by a pair of cams carried on a driving shaft supported on the main frame. At the end of the screen a steel bumping angle is rigidly fixed, which strikes a bumping beam

secured to the frame and limits the forward movement of the screen. The amount of movement is
adjustable.

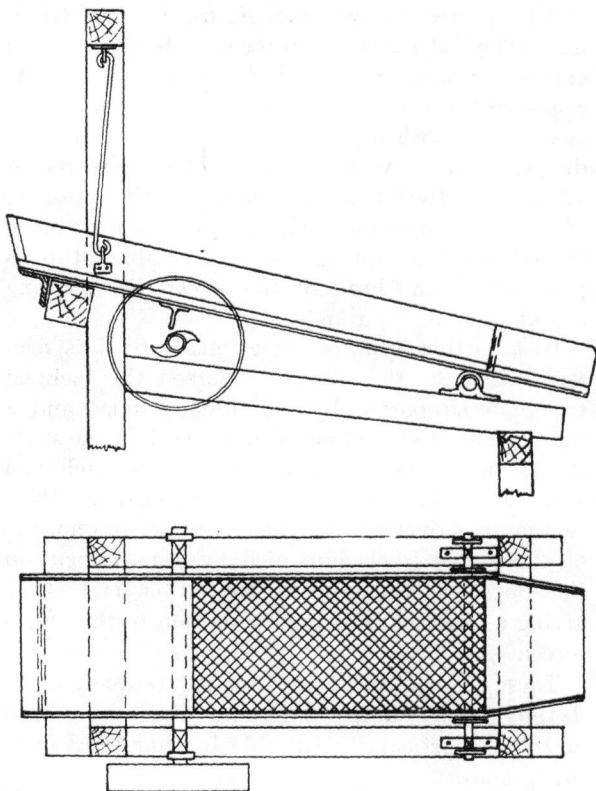

FIG. 7.—Shaking Screen.

The arrangement of the screen is shown in
Fig. 7.
2

In a modified pattern of this screen a steel bracket is fixed underneath the trough, to which is secured a long rod running parallel to the screen and supported in two brackets fixed to the framing. The rod carries a stop collar, between which and the lower bracket is a strong spring. The upper end of the rod carries a roller which presses against a toothed disc or cam carried on the driving shaft. As the cam revolves, the rod is alternately forced forward against the pressure of the spring and released, the return stroke being caused by the spring. An adjustable stop is provided at the head of the screen, for varying the amount of " bump."

In a further type of screen made by this firm and known as the " Vibro " Screen the inclined trough is supported by four hinged links, and a rapid transverse motion is given to it by a shaft arranged longitudinally and having at each end a small crank. The cranks are opposed at 180°, so that a movement in one direction at one end of the screen is accompanied by a movement in the opposite direction at the other end, thus giving a twisting or wriggling motion to the whole screen.

These screens have proved very successful in dealing with damp materials, and the Vibro Screen is particularly suitable for handling very fine grades.

Fig. 8 shows the arrangement of a typical Colliery Jigging Screen, on to which the coal is tipped direct from the tubs. The large coal passing over the screen is delivered on to a picking belt of the Tray Conveyor type, while the small coal is discharged on to a separate conveyor, by which it is delivered to the washery.

The screen shown in the figure was constructed by Messrs. M. Coulson & Co., Ltd.

The Spiral Separator, designed specially for the elimination of " dirt " from coal, is installed in this country by Messrs. Hugh Wood & Co., Ltd., of Newcastle, and operates by taking advantage of the difference in the coefficient of friction between the coal and the stone and bastard coal or shale forming the greater part of the impurities, when sliding down an inclined shoot formed of iron or steel. The coal has a lower coefficient of friction than the stone, while the bastard coal has a coefficient somewhere between that of the coal and the stone.

The inclined shoots are arranged in the form of spirals, supported from a central vertical post, the material passing down the shoots by gravity. There are three separating shoots of comparatively narrow width, surrounded by an outer spiral of steel plate, in which the coal is collected, the bottom of the separating shoots being formed with perforations and ridges or corrugations of special shape to assist the friction in separating the various grades.

The raw coal is fed into the separator by a shaker conveyor or screen giving a regular feed, and is divided by the feed shoots into three streams which enter the three separating spirals. It is important that the coal should be screened to an even size before being fed to the separator. Coal

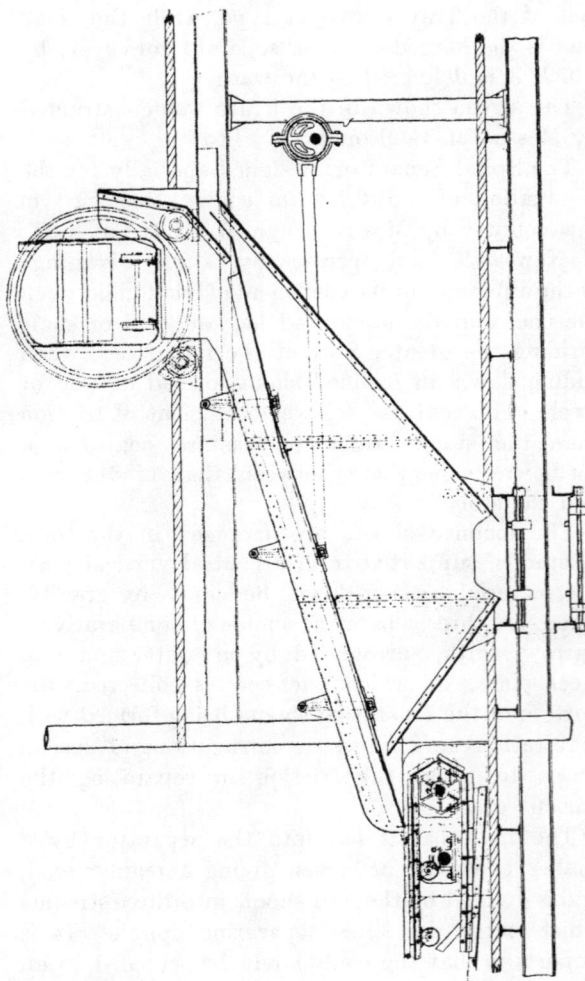

FIG. 8.—Jigging Screen.

from 4 in. to $\frac{3}{8}$ in. can be dealt with, but it should
be divided into the following approximate sizes :—

Through 4   in. and over $2\frac{1}{2}$ in. mesh.
,, $2\frac{1}{2}$,,   ,,   ,, $1\frac{1}{2}$,,   ,,
,, $1\frac{1}{2}$,,   ,,   ,, $\frac{3}{4}$,,   ,,
,, $\frac{3}{4}$,,   ,,   ,, $\frac{3}{8}$,,   ,,

The feeder should be arranged to take out any
fine material that may have been made between
the sizing screens and the separator, and it should
also ensure that the coal enters with the minimum
drop in order to avoid any further breakage, and
that the coal is delivered on to the separator
without any initial velocity.

As the raw coal slides down the separator the
velocity gradually increases up to a point where
the centrifugal force on the coal overcomes the
friction. The coal then moves outwards and falls
over the edges of the separating spirals into the
outer shoot, while the stone with its higher coefficient of friction remains in the separating threads
and is discharged at the bottom into a separate
pocket. The bastard coal, which has an intermediate coefficient of friction, keeps to the outer
edges of the stone threads and can be collected
separately, if required, for burning under boilers.

The usual capacity of each separator, when
dealing with sized bituminous coal, is six to eight
tons per hour, and the feeder should be arranged
to supply this amount regularly. Each separator
is designed specially to suit the particular class
and size of coal to be handled.

Spiral Separators have for some years been

largely used in America, and a number of plants have more recently been installed in this country.

No power, beyond a small amount for driving the feeder, is required, the whole of the work being done by centrifugal force and gravity.

Various tests have shown that coal can be cleaned

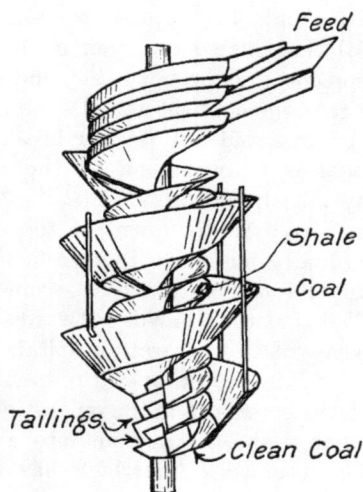

FIG. 9.—Spiral Separator.

by these separators until only 1 to 2 per cent. of slate and similar impurities is left in the coal.

As a general rule, the coefficient of friction for shale is approximately 50 per cent. greater than that for coal, the coefficient for grains of coal being about 0·4 and for grains of shale about 0·6.

The arrangement of a typical Spiral Separator is shown diagrammatically in Fig. 9.

CHAPTER II

An important part of the operations at a colliery pithead is the removal from the coal of pieces of shale, rock and other foreign material which may be brought up from the pit with the coal. The removal of this " dirt " in the case of the larger grades of coal is generally carried out by hand on Picking Belts or Tables. The coal which passes over the preliminary screens, on which the smaller sizes are removed for treatment in other ways, is delivered by inclined shoots on to the tables.

These tables consist of conveyors, either of the belt, steel tray or jigging type, on either side of which run platforms for the pickers. The belts or trays have a width of about 4 ft. 6 in., which allows the whole area to be covered by the two rows of pickers, and they travel at a speed of 40 to 50 ft. per minute.

Belt conveyors are of the ordinary type, usually with heavy canvas belts, carried on horizontal rollers. Tray conveyors consist of steel plates or slats supported by endless chains having links about 12 in. pitch, the plates overlapping slightly so as to form a continuous travelling table.

The usual length is 60 to 75 ft., the longer length being used when the delivery end section is arranged to be lowered into the railway wagons to avoid breakage.

The capacity of a picking belt 4 ft. 6 in. wide is about 60 tons per hour. When used as a conveyor only the speed may be increased to 60 ft.

per minute and the capacity is then about 80 tons per hour.

From the ends of the picking belts the coal is usually delivered direct into railway wagons over shoots, which sometimes contain an auxiliary screen for removing any dust or small coal which may have been produced subsequent to the preliminary screening.

It is important that the screened and picked coal should be lowered as gently as possible into the wagons, otherwise the screening which has already been carried out will be partly nullified by the breakage which will occur.

Instead of a shoot a lowering conveyor may be used, the conveyor chain being provided with crossbars at suitable intervals. These bars hold back the coal and prevent it sliding freely down the shoot, which can be raised gradually as the wagon fills up.

In other cases, the end section of the picking belt is itself arranged on a hinged jib, which may be raised and lowered to suit the height of coal in the wagons.

The Zimmer and similar allied types of reciprocating or jigging conveyors are also used as picking belts, for which purposes they have the advantage that the reciprocating motion spreads out the coal automatically into a thin layer, whereas in the ordinary pattern of belt the coal is deposited in heaps which must be spread out before the impurities can be seen and removed.

Circular Picking Tables are made by Messrs. Fraser and Chalmers, Ltd., for cleaning ore and removing pieces of rock containing little or no valuable mineral, or for separating ore sufficiently rich to be shipped direct without concentrating from that which requires further treatment.

These tables consist of annular rings 20 to 30 ft. in diam., carried on rollers and revolved by gearing. The pickers can stand both inside the ring and outside, and the tables can, if necessary, be made with a double deck. The ore is fed on to the table at one point through a rotary screen which removes the material which is too small for hand picking. The ore to be picked is carried round the table for nearly a complete circle, and is then removed by a plough and discharged over a shoot for any further treatment required.

If washed material is to be dealt with the table may be perforated in order to drain away surplus moisture.

Conveyor Screens

In many cases it is convenient to combine the screening and conveying of the material, the larger pieces being discharged at the end of the conveyor, while the smaller pieces are deposited at intermediate points or discharged on to separate conveyors leading them to their respective bins or hoppers.

Belt conveyors made of woven wire are occasionally used for this purpose, especially for draining water from washed material, or an ordinary crossbar conveyor may be used with screen bars placed in the bottom of the trough, over which the material is dragged by the chain, but in the

2*

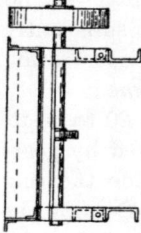

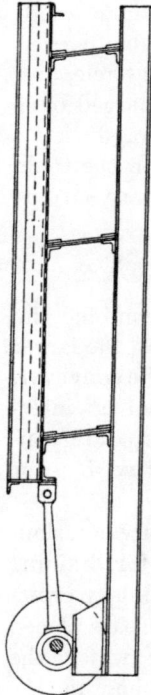

Fig. 10.—Zimmer Screen Conveyor.

majority of cases a conveyor of the jigging or reciprocating type is used.

This type of conveyor forms a very efficient screen for many materials, if the bottom plates are perforated with holes of suitable size, as owing to the way in which the material is spread out by the motion of the conveyor, a comparatively short length of conveyor screen is required.

The jigging conveyor consists essentially of a long rectangular trough reciprocated rapidly by means of a crank and connecting rod. The trough is supported by strips of ash placed at a slight angle or by rollers working in curved paths in such a way that at every forward stroke the trough has a slight upward motion. The combination of horizontal and vertical motion causes the material to be thrown forward and off the surface of the bottom plate at every stroke, so that it proceeds forward along the conveyor in a series of small jumps at rapid intervals.

A number of screens may be incorporated in one conveyor, commencing with the smallest apertures and gradually increasing the size of the openings, and in this way the various sizes of the material can easily be separated out and deposited in separate bins or wagons.

Jigging conveyors are particularly suitable for dealing with coal and similar material, but the author has found that when handling coke or other material which may contain long tapering pieces there is a tendency for these pieces to lodge in the holes, thus causing an obstruction to the flow of material along the conveyor which the jigging motion is not able to overcome entirely. Some improvement is effected by bending down, instead of removing completely, the pieces of metal which are cut away to form the holes, as these tongues tend to give any pieces lodging in the holes a kick at each stroke.

The general arrangement of a Zimmer Screen is shown in Fig. 10.

G. F. Zimmer gives the following figures for the capacity in tons per hour of a vibrating conveyor of his design when handling coal and coke :—

Depth of trough in inches.	Width of trough in inches.							
	12	16	20	24	36	48	60	72
Coal :								
4	6– 7	8– 9	10–12	13–15	—	—	—	—
6	9–10	13–15	16–18	18–20	30–32	35–40	45–50	50–60
8	—	—	—	25–30	35–40	50–60	60–70	70–80
Coke :								
4	$3\frac{1}{2}$–4	5–6	6–8	8–10	10–13	16–19	19–22	22–26
6	—	—	—	12–14	16–19	24–28	28–33	33–39
8	—	—	—	16–19	22–26	33–39	39–46	46–53

The figures given above are for a horizontal conveyor, but this type of conveyor will work at a slight incline either up or down. The average speed at which the material is conveyed is about 50 ft. per minute. Mr. Zimmer also states that a conveyor of this type 100 ft. long carrying a load of 50 tons requires to drive it 8·75 H.P.

When handling coal similar figures to those given above may be taken for the capacity of screening conveyors, but when dealing with coke lower figures should be taken unless it is possible to keep a man in attendance to make sure that the action of the conveyor is not impeded by pieces of the material lodging in the holes.

Jigging screens of the Zimmer or Marcus type are sometimes made in double deck form with the two troughs moving in opposite directions simultaneously. This enables the balancing of the screen to be carried out much more completely than in the single deck pattern, and thus prevents excessive vibration in the supports.

The driving shafts revolve at 78 to 64 r.p.m., and the travel of the screens is 8 in. to 12 in., according to size, giving a speed of 45 to 60 ft. per minute for the material.

The " Marcus " Horizontal Screen Conveyor, made by Messrs. Head, Wrightson & Co., Ltd., of Stockton-on-Tees, is of the reciprocating type, and consists of a steel trough supported on rollers and propelled forwards and backwards by means of cranks and links. The driving mechanism is so constructed that a variable speed is obtained during each stroke, the forward stroke commencing slowly, with a uniform acceleration up to about three-quarters of the stroke, and a quick retardation during the last quarter. On the return stroke, a rapid acceleration is given during the first quarter, with a uniform retardation during the remainder of the stroke.

This differential motion results in the material being carried along a short distance at each forward stroke, and owing to the momentum given to it, the material is left behind on the return stroke, so causing it to travel along the screen. This result is obtained by means of short drag links connecting cranks on the driving and propulsion shafts, which are arranged slightly out of line with one another, the propulsion shaft being connected to the troughs by means of eccentrics and connecting rods.

In the latest pattern of Marcus Screens, double troughs are so arranged that the shock effect obtained with one screen is neutralised by that from the second screen, thus preventing any vibration and obviating the necessity for heavy and expensive supports.

The arrangement of a typical Balanced Marcus Propulsion System is shown in Fig. 11, the two screens being designed to convey the material in the same direction simultaneously, but the gear can if required be designed to convey the material in opposite directions.

If fitted with a solid bottom plate instead of screen plates, the Marcus Conveyor can be used as a picking table or as an ordinary conveyor.

The tables below give the capacities of various

sizes of Marcus conveyors, but the driving motors should be capable of developing twice the full load power given in the tables in order to overcome the additional power required for starting.

Type.	Size.	Stroke in inches.	Speed r.p.m.	Weight of both troughs. Cwts.	Weight of material in both troughs. Cwts.	Approx. full load. H.P.	Approx. conveying speeds. Ft. per min.
A	1 A	8	78	40	40	5	39
	2 A	10	72	60	60	9	45
	3 A	12	64	80	80	$12\frac{1}{2}$	48
	4 A	12	64	100	100	$15\frac{1}{2}$	48
	5 A	12	64	120	120	19	48
B	1 B	8	78	80	80	10	39
	2 B	10	72	120	120	$17\frac{1}{2}$	45
	3 B	12	64	160	160	25	48
	4 B	12	64	200	200	31	48
	5 B	12	64	240	240	38	48

In type A the two screens are arranged for propelling the material in opposite directions, while in type B the material in both troughs is propelled in the same direction simultaneously.

The differential motion given by the Marcus propulsion gear results in the coal or other material being spread out in a thin layer over the surface of the trough, thus greatly assisting the efficiency of the screening, or when used as a picking table, bringing all the shale into full view of the pickers.

A screen of the Zimmer type, made by Messrs. The Coal and Ore Dressing Appliances, Ltd., of Westminster, is supported on double legs of ash and is reciprocated by means of an adjustable eccentric, so that the stroke may be varied to suit the nature of the material. By this means the screen can be made to deal successfully with coarse or fine material, either wet or dry.

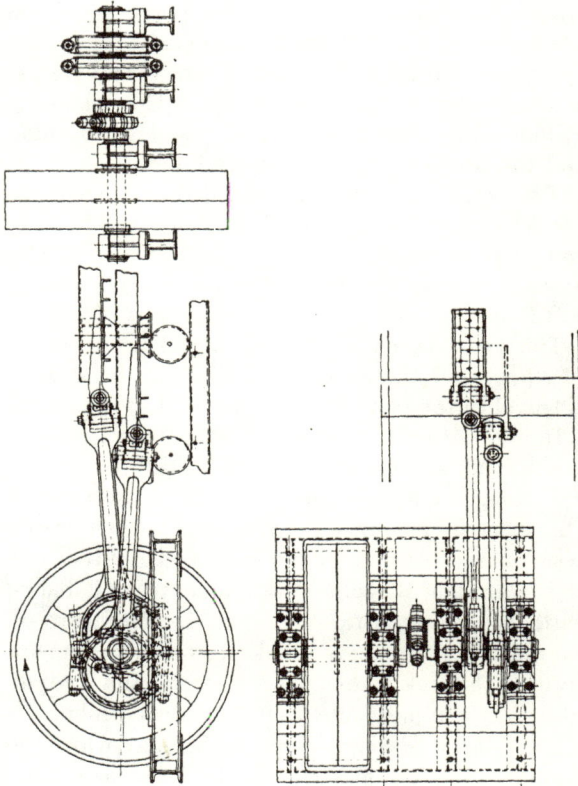

Fig. 11.—Arrangement of Balanced Marcus Gear.

The Norton Jigging Screen is generally similar in construction to the Zimmer Screen Conveyor, 'but the trough is carried on inclined links attached to the underside of the trough and to the supporting structure. The stroke is longer than in the Zimmer Screen and the speed is less. The trough is usually made in two portions, which are arranged to reciprocate in opposite directions by means of links and rocking levers. This results in practically complete balance, and air-cushioning cylinders are also fitted to ensure smooth running and to reduce the power consumption.

Two screens of this type installed at Tottenham Gas Works for handling coke are each 105 ft. long and have a capacity of 50 tons per hour. The power required for driving each screen is about 5 B.H.P.

The usual speed of the driving shaft is about 90 r.p.m. and the stroke 5 to 6 in., giving a rate of travel to the material of about 40 ft. per minute.

In another pattern of screen the vertical motion is obtained by means of curved guides secured to the trough and to the supporting framework. The trough is carried by rollers which work between the upper and lower sets of guides, and as the trough is reciprocated forward and backward by the eccentric or crank, the curvature of the guides gives it a quick upward movement towards the end of the outward stroke and a rapid fall at the commencement of the return stroke.

The Harris Patent Vibrator Screen, which is manufactured by Messrs. W. J. Jenkins & Co., Ltd., Retford, has been designed specially for handling and screening coke and for obtaining a number of different grades within a limited space. In order to overcome the difficulty sometimes experienced with the ordinary type of jigging screen when dealing with coke, by reason of the material hanging up in the holes, a somewhat novel arrangement has been adopted by means of which the screens are given a vertical movement in addition to the vibrating or swinging motion.

As will be seen from the illustration (Fig. 12), the Harris Screen consists essentially of a framework carrying two inclined pans or troughs of the usual jigging type, supported from the framework by links. The pans slope in opposite directions and are reciprocated by means of eccentrics mounted on a common shaft.

Inside the two pans double screens are carried by means of springs at their upper ends attached to brackets supported from the sides of the troughs. The reciprocating motion of the pans is transmitted to the screens, which at the same time move vertically on account of their spring suspension. The movement of the screens relative to the carrier pans is limited by adjustable stops, but is of sufficient amplitude to ensure the coke rising from the screen and to prevent the holes from becoming clogged.

The double motion of the screen ensures rapid and efficient separation of the various sizes, so that it is possible to arrange a screen of large capacity in a comparatively small space, and to obtain a large number of different grades of coke from one plant. With double screens of different mesh in each of the two carrier pans, five sizes of coke are obtained.

The Harris Screen is suitable either for loading the coke direct into bogies or trucks, or into

hoppers for filling carts and bags. Actual tests
on plants in daily use at a large gas works have

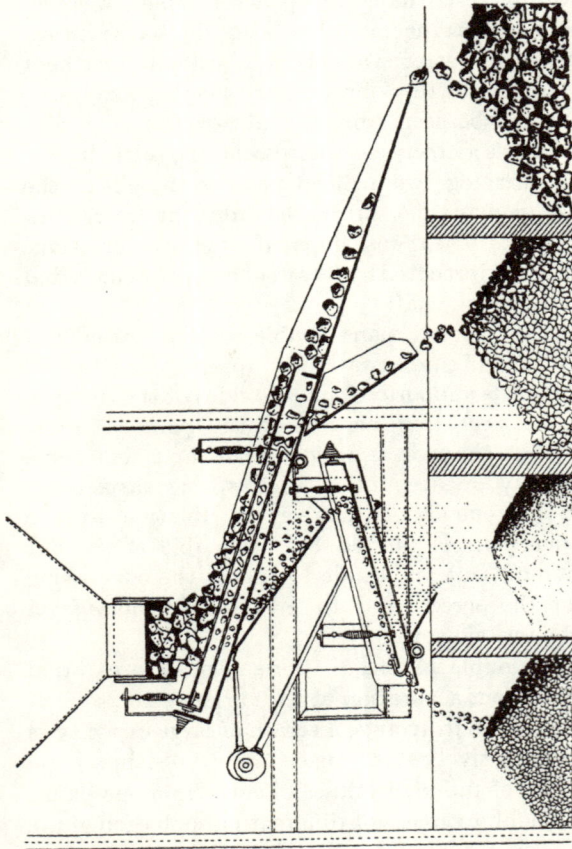

FIG. 12.—Harris Vibrator Screen.

given the following results when working on
ordinary gas coke :—

Large coke 58 per cent.

2 in. ,, 29 per cent.

$\frac{3}{4}$ in. ,, 7 per cent.

Breeze 6 per cent.

When dealing with 35 to 40 tons of coke per hour, the power absorbed is about $5\frac{1}{2}$ B.H.P. and the screen runs at a speed of 125 r.p.m.

Where screening is carried out by means of link belt or similar conveyors during its travel from one point of the works to another, it is only possible to separate the material into two sizes : the small, which passes through the belt, and the remainder, which is delivered over the end of the conveyor.

A shoot must be arranged between the two strands of the chain so that the fine material passing through the upper strand falls clear of the bottom one.

With this type of conveyor screen the material is subjected to little or no agitation during its travel and it must therefore be delivered on to the belt in a very thin layer, otherwise a large proportion of fine material will be carried forward and delivered with the large at the end of the conveyor. Instead of the link belt, two strands of chain may be used, supporting between them a series of trays made of perforated plate or woven wire.

This type of conveyor screen is sometimes used for de-watering or draining coal and other materials after washing, and it is also used as a feeder to crushing rolls in order to remove the smalls and so reduce the work of the crusher.

CHAPTER III

THE Rotary Screen, sometimes known as a Trommel, forms a very efficient means of separating out the various sizes of a material into separate bins or hoppers, as the rotating motion of the screen effectively agitates the mass of material during its travel along the screen and thoroughly separates the small from the large particles.

Rotary screens are largely used for handling coke, gravel, pan breeze, stone, clinkers and similar material, but they are not altogether suitable for soft materials owing to the breakage caused by the motion of the material as the screen revolves.

The power required for operating a rotary screen is comparatively low. The screen sections may be constructed of steel plate, perforated with round or square holes, woven-wire mesh or expanded metal, according to the nature of the material to be handled.

The screen plates are attached to a rigid cylindrical framework, carried on rollers or a central shaft, and rotated by gearing. The axis of the cylinder is usually inclined slightly to the horizontal, suitable means being provided for taking the end thrust caused by the inclination of the axis, so that the rotating motion of the screen carries the material slightly forward at each revolution, while at the same time it rolls over the perforated plates. In some cases the screen is made in the shape of a truncated cone, with the axis horizontal, and this arrangement, though slightly increasing the initial cost, has the advantage that it simplifies the driving arrangements and eliminates the end thrust which is necessarily present in the inclined pattern.

Usually each screen is provided with several rings of plate having different perforations arranged in order of size, the smallest naturally being at the inlet end. Shoots or hoppers are fixed under each section for the various sizes of material. The first section is often made with a double casing, the outer screen being of finer mesh than the inner plate, in order to separate out the finest grade of material. This arrangement reduced the amount of work done by the first part of the screen, and is particularly useful when the material contains a large proportion of smalls.

The lower end of the cylinder is left open, so that the tailings or material which is larger than the openings in any of the screen sections are delivered to a separate bin or heap. The upper end is partly closed by a plate or cast-iron ring so that the material cannot fall out at this end, and a shoot is fixed for feeding the material in through the annular space which is left open.

At each revolution the material is carried more or less round the circumference of the screen, according to the speed at which it revolves, and it then falls or rolls on to a fresh portion of the screen plate. This motion results in very effective screening, but as mentioned above may cause a considerable amount of breakage in soft materials. For this reason, rotary screens are not much employed for colliery work.

When the screen is supported on rollers the

drive is usually effected by means of an annular spur or bevel ring mounted in a convenient position round the drum, while in the other type, in which the drum is carried by spiders from a central shaft, a bevel gear drive is arranged at the lower end, with a tail bearing designed to take a certain amount of thrust fixed between the gear and the end of the drum.

The greater speed of the material at the larger end of the screen facilitates the separation of the large and small particles.

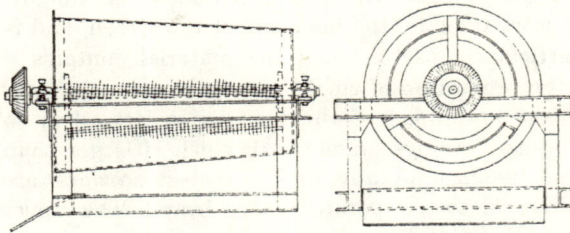

Fig. 13.—Taper Rotary Screen.

The arrangement of a conical screen of this type, as made by Messrs. W. J. Jenkins & Co., Ltd., of Retford, is shown in Fig. 13.

In cement works, a type of rotary screen known as a Reel is often used for separating out large particles which have not been reduced to the required size in the grinding mills. The cylinder is surrounded by a casing along the bottom of which a screw conveyor is arranged for removing continuously the material which passes through the screen plates, while the large particles pass over the end of the screen and are returned by a shoot or a separate conveyor to the grinding mill.

In some cases a tapping or shaking device is incorporated, in order to dislodge any of the fine material which adheres to the screen, but the noise caused by the hammer blows is an objection to this arrangement.

The usual inclination of a rotary screen is about 1 in 8 to 1 in 10 and the peripheral speed 200 ft. to 250 ft. per minute, or, say, 15 to 20 r.p.m. for a screen not exceeding 4 ft. diam.

The capacity of a rotary screen depends on the physical nature of the material and its condition as regards moisture content, dry material screening more readily than wet, and on its specific gravity. It also depends greatly on the kind of material of which the screening surface is constructed, the proportion of the openings in relation to the total area, the shape of the screen and the rate of travel.

The rapidity of screening has been found to vary approximately in proportion to the square root of the diameter of the holes, the weight of material passing through the screen per hour per square foot of screening surface being given by the formula

$$w = x\sqrt{n}$$

where x is a constant which must be determined by experience for any particular class of material, and

n is the diameter of the holes in sixteenths of an inch, or if a square hole is used, the diameter of a circular hole of equivalent area.

The following figures were recently published *
for the output of rotary screens dealing with
siliceous copper sulphide ores. In one case a
parallel cylindrical screen was used, 50 in. diam.,
10 ft. long, constructed of $\frac{3}{16}$ in. plate, the first
6 ft. perforated with holes $\frac{1}{4}$ in. diam. and the
remaining 4 ft. with 1 in. holes. The speed was
16 r.p.m., and the inclination 1 in. per foot.

Capacity—

through $\frac{1}{4}$ in. holes	4·1	tons per hr.	=	7·3	lb. per sq. ft. per hr.
„ 1 in. „	5·4	„	„	= 14·4	lb. per sq. ft. per hr.
over screen	9·7	„	„		
Total	19·2	„	„		

In the second case a conical screen was used,
tapering from 48 in. to 60 in. diam., 14 ft. long,
running at 15 r.p.m. Slope $\frac{7}{8}$ in. per ft. Screen
plates $\frac{3}{16}$ in. thick, the first 7 ft. perforated with
$\frac{3}{8}$ in. holes and the remainder with $\frac{3}{4}$ in. diam.
holes.

Capacity—

through $\frac{3}{8}$ in. holes	5·5	tons per hr.	=	8·8	lb. per sq. ft. per hr.
„ $\frac{3}{4}$ in. „	8·6	„	„	= 12·3	lb. per sq. ft. per hr.
over screen	11·2	„	„		
Total	25·3	„	„		

For this ore the factor x in the formula given
above is about 3·56.

As mentioned above, the capacity is governed
by various conditions, and the factor will vary
considerably according to the type of screen surface
and the nature of the material handled.

* *Mechanical World,* April 11th, 1924.

By using square-mesh woven-wire screens instead of perforated plate, the capacity for a given length and diameter of screen is greatly increased on account of the greater proportion of the total area formed by the holes, but the cost of maintenance will be increased considerably, as the wear and tear of such a screen is nearly double that of a perforated plate screen having holes of equivalent size.

The circular screen is the most common pattern, but in some cases the framework is made polygonal, and this form has certain advantages. The screen plates are flat and are therefore more easily replaced when worn, and the angular shape helps to break up the mass of material and so slightly increases the capacity and efficiency of the screen. For this reason, the circular form is preferable if the material is at all friable and undue breakage is to be avoided.

A few makers suspend the cylinder by chains or belts from pulleys fixed to a driving shaft supported over the screen, the chains also transmitting the motion to the screen.

Bars are sometimes used for heavy work, but the screen sections are generally made of perforated steel plate for screening materials up to 4 in. diam., or wire gauze for fine materials up to about $\frac{1}{8}$ in. diam. Although revolving screens have the greatest capacity of the various types in common use, they are not suitable for very fine materials smaller than 20 mesh.

Double screens are frequently used, the main cylinder being surrounded for a portion of its length by a second screen perforated with smaller holes. The material passing through the inner screen is separated into two or more grades, the larger pieces which are rejected by the outer screen passing out through the annular space between the two cylinders into a separate bin or shoot. This arrangement is particularly suitable for screening material which contains a large proportion of smalls, as the grading is much more efficient than with a single cylinder, unless the latter is of excessive length or the rate of feed unusually low.

This arrangement may be further modified by using three or four concentric cylinders, each cylinder being perforated with one size of hole only. In this way the strongest screen handles the greatest amount of material, thus increasing the life of the screen and reducing the overall length required for efficient grading.

For ordinary work the length of each section of a rotary screen should be about 5 to 6 ft., but this will depend to some extent on the proportions of the various sizes present in the material to be handled.

The " New Century " Rotary Screen, made by Messrs. Fraser and Chalmers, Ltd., is designed particularly for the preliminary treatment of materials embedded in clay, mud, loam or soapstone, and is used for washing and grading ore, sand and gravel. The drum, which is fitted externally with screen plates of any required mesh, has a series of 1 inch bars fixed internally spaced 1 inch apart and extending the whole length of the screen. As the drum revolves, these bars break up the lumps of material, separating the clay from the gravel or rock, and they also serve to protect the screen plates from damage by the large lumps.

The fine material passes between the bars and through the perforations of the screen.

The drum revolves in a trough of water, so that as the material is broken up it is washed free of the adherent clay, which is deposited, together with the fine portion of the rock or gravel which has passed through the screen, in the tank, from which it is picked up and discharged into a launder for further separation, by means of a series of buckets attached to the outside of the drum. The large pieces of material, which are retained on the bars, travel along the screen, and are thoroughly washed during their progress. They are then picked up by a plough-shaped chute and discharged through the end of the screen.

Messrs. Fraser and Chalmers also make a rotary screen for dealing with fine materials in the shape of a cone with its axis inclined at an angle of about 45° and driven through bevel gears. The sides of the cone are inclined at the same angle, so that as the screen revolves one part is horizontal and the other side is approximately vertical. A casing is arranged round the revolving cone, into which the material passing through the screen is discharged and from which it is led away through an outlet at the bottom, while the oversize material which is retained on the screen is discharged through the hollow spindle on which the screen revolves.

To assist in keeping the screen clean a perforated spray pipe is arranged along the vertical side of the screen.

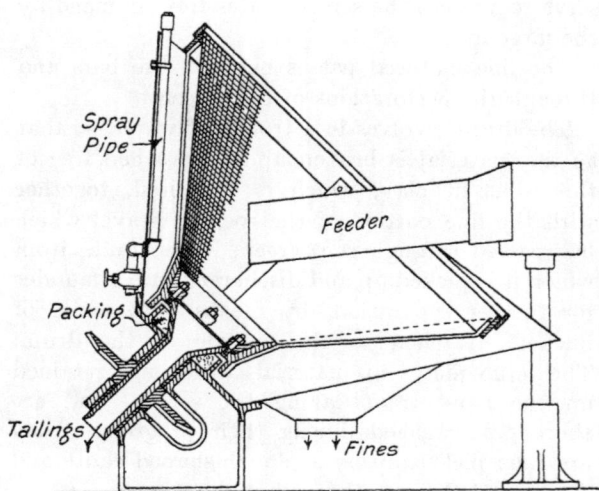

Fig. 14.—Conical Rotary Screen.

A feed distributor, supported on an external post, ensures a regular and even feed on to the screen.

The arrangement of this screen is shown in Fig. 14.

CHAPTER IV

THE preparation of many materials, such as coal, sands and ores, calls for the removal of foreign matter or " dirt," of various kinds, and in many cases this dirt can best be removed by treating the material with water in one or other of the many types of washers now in use.

The selection of the most suitable washer for any particular case depends on the nature of the impurities to be removed and the relation between the specific gravity of these impurities and that of the material from which they are to be separated, and also on the presence or absence of an ample supply of washing water.

In dealing with sands of various kinds, the principal impurity met with is clay, which can readily be removed by washing. Where there is a plentiful supply of water and a comparatively small quantity of material to be dealt with, an ordinary trough washer may be used. This type of washer consists of a long trough, generally arranged at a slight incline, into which the sand and water are fed. The sand is agitated either by hand or by mechanical means. The clay is washed out and carried along in suspension by the water while the sand is left behind in the trough, from which it is removed at intervals by hand.

If it is desired to separate sand from gravel in this form of washer, screens must be incorporated in the upper part of the trough, through which the water and sand can pass, while the larger gravel is retained on the screens.

When sand is contaminated by iron, although some of the iron compound is removed by the water, the greater part is left behind with the sand, and other methods must be adopted to ensure the thorough elimination of the iron. For many purposes the complete removal of the iron compounds is not necessary, but for glass making and in other cases where the presence of iron would cause discoloration of the finished product the removal of most of this impurity is essential.

The flow of water in this type of apparatus must be adjusted until it removes the greatest possible proportion of the impurities without carrying off any considerable amount of the finest sand which it is desired to retain.

The washed sand is discharged into an elevator having perforated buckets, in order to allow as much of the water as possible to drain away.

In some cases the material is carried through the washer in a direction opposite to that of the flow of the water by means of a kind of worm conveyor consisting of blades or paddles set at an incline round a revolving shaft. In this way the cleanest sand meets the cleanest water and the efficiency of the apparatus is considerably improved.

A further modification of this type of washer consists in using a closed trough in the form of a tube, either horizontal or set at a slight incline, and this arrangement has the advantage of requiring considerably less water than the open trough, while the washing is under better control. The power required for operating the washer is approximately $\frac{3}{4}$ H.P. per ton of material washed per hour, the larger machines requiring rather less power in proportion than those of smaller capacity.

In some cases, as in the Blackett Washer, the drum is revolved in a similar manner to a rotary screen, and in this type screen plates are frequently arranged at the delivery end so that the washed material is separated into various sizes.

The smaller sizes of coal separated out by the preliminary screening plant, and particularly that portion which is destined for the manufacture of coke, are usually cleaned by washing to remove as much as possible of the dirt and other impurities with which the coal is always unavoidably contaminated.

By this means grades of coal which would otherwise be practically useless can be utilised successfully for the manufacture of high-class coke, and can also be burnt under boilers quite satisfactorily.

The operation is facilitated by the fact that the shale, pyrites, etc., forming the greater part of the impurities are heavier than the coal, though there is always a certain proportion of the foreign matter which is actually embedded in the coal and cannot be separated by any mechanical means.

The principle employed in the various types of coal washers is the agitation of the coal in a plentiful supply of water, which is generally available from the pit itself.   In this way the impurities are separated out and fall to the bottom of the washer, while the coal is carried forward over screens where the surplus water is drained out.

One type of washer consists of a long inclined trough about 3 ft. wide and 12 to 15 in. deep. The length may vary up to 100 ft. The coal and water are introduced at the upper end of the trough, and the necessary agitation is produced either by crossbars fixed to the bottom of the trough, over which the stream of water and coal falls, or by an endless chain carrying crossbars or scrapers, which moves against the flow of the stream.

In the one case the heavier impurities lodge in the compartments formed by the various crossbars, from which they are removed at suitable intervals, and in the other type the heavy materials are carried up to the top of the incline by the chain and are then dumped into wagons.

In the Blackett Coal Washer and similar types of machine, the agitation is produced by the rotary motion of the drum, and a steel spiral is fixed inside the drum in such a way that the dirt which lodges against it is carried up the incline against the stream of coal and water, the washed coal being delivered at the lower end on to the usual draining screen.

Fig. 15 shows the arrangement of a Blackett Washer as made by Messrs. M. Coulson & Co., Ltd., of Spennymoor. In a recent installation three washers are installed, each 4 ft. diam. by 30 ft. long. The coal is screened before washing into three sizes, peas, nuts and smudge, and is treated separately. The inclination of the drum is adjusted to suit the class of coal, and the speed is regulated according to the amount of dirt present with the coal.

In other cases mechanical stirrers are used, but the general system of operation remains the same.

The " British Baum " Washer, made by Messrs. Simon - Carves, Ltd., of Manchester, is very largely used for coal washing, and is typical of the type of apparatus in which a closed vessel is used and the agitation of the water and coal produced by mechanical means.

In the Baum Washer compressed air is used for agitation, the blasts of air being admitted at intervals by means of piston valves operated by eccentrics from a revolving shaft arranged over the apparatus. The pistons make 40 to 50 strokes per minute. The washer box consists of a long steel-plate vessel or tank having a semi-circular bottom and a vertical diaphragm running the full length of the tank and extending rather more than half the depth. On one

3

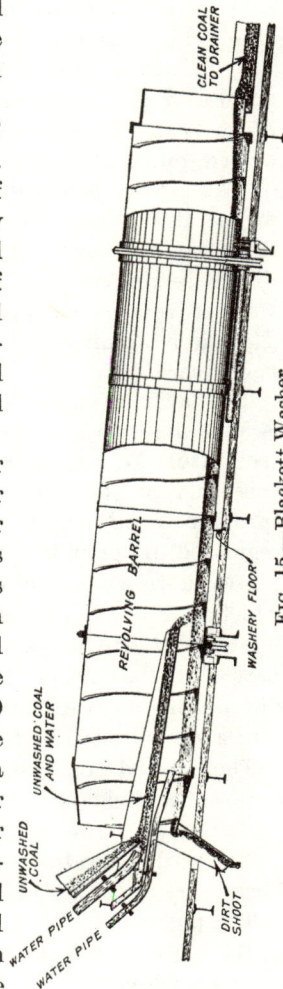

FIG. 15.—Blackett Washer.

side of this diaphragm the tank is completely closed in, thus forming an air chamber, on the top of which are mounted the valves for the admission of the compressed air. The other side of the tank is provided with a perforated grid or bed, on to which the coal is fed. The agitation of the water produced by the pressure of the air, as it is alternately admitted and exhausted, lifts the coal and agitates it thoroughly. The dirt is separated from the coal during its downward movement when the exhaust port is opened. The air is admitted at a pressure of about 2 lb. per square inch.

The raw coal enters the washer at one end, and is deposited on the perforated bed in the first section of the box. The heavy dirt collecting on this part of the bed is drawn off continuously through an adjustable sluice gate, which can be regulated to give the desired rate of flow, and passes down a shoot into an elevator by means of which it is removed and drained. The coal and lighter dirt are carried forward into the second part of the box, where the process is repeated, the dirt passing out through a second adjustable sluice into a second elevator for drainage and removal.

The washed coal passes out through a further outlet at a slightly higher level and is taken off to the classifying screens or bunkers.

The washing water enters through valves at the lower part of the air chamber.

The fine dirt which passes through the openings in the perforated bed plates collects in the bottom of the washing box, and is removed by means of

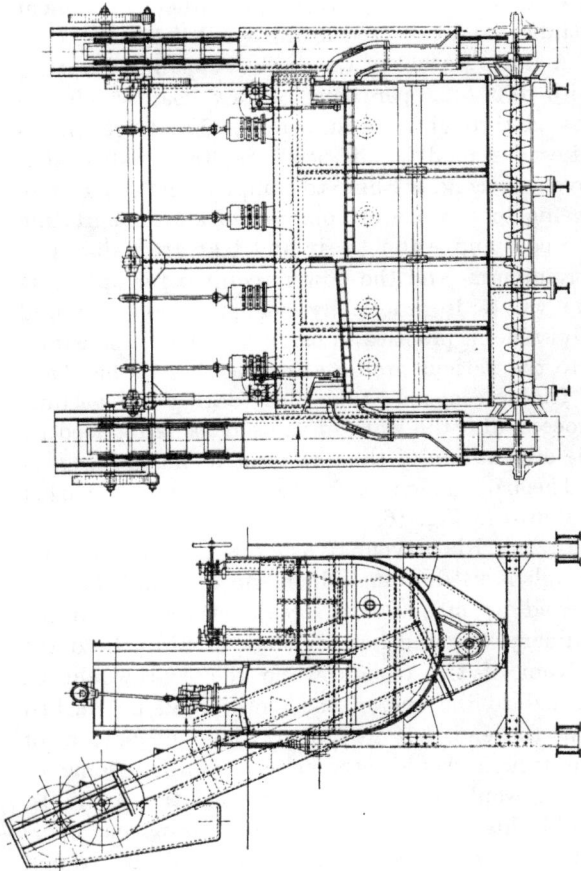

FIG. 16.—British Baum Coal Washer.

worm conveyors arranged in a subsidiary chamber attached to the box.

The water passing out with the coal is drained

out at the screens or in any other convenient manner, and collected in the settling tank.

A single washer of this type is capable of dealing with 100 tons per hour of raw coal which has passed through a $3\frac{1}{2}$ in. screen. For larger quan- tities a second box is used and the small coal is given a second washing to complete the separation. Owing to the use of compressed air for agitating the coal and water, wear and tear are reduced to a minimum, and the power required for operating the valves is practically negligible. It is found advisable in practice to carry out the classification into the various grades after washing, as breakage of friable coal may take place during the washing process, also the work of the screens is reduced by the elimination of the dirt prior to the grading.

The arrangement of the " British Baum " Washer is shown in Fig. 16.

In the Rheolaveur Washer a long rectangular trough is used, along which the coal and dirt are carried by means of a stream of water. In the ordinary way the heavier shale would fall to the bottom of the trough while the coal would be carried to the end of the trough and be washed over the end, but in a short time the bottom of the trough would become full of dirt, some of which would be carried over with the coal. To avoid this contamination of the clean coal aper- tures are provided in the bottom of the trough through which the shale is removed, but to prevent the rush of water through the openings carrying with it some of the coal, the apertures are each fitted with the Rheolaveur apparatus. This consists of a cast-iron box provided with a central diaphragm and having one opening at the top corresponding to the slot in the bottom of the trough, and a second aperture at the bottom provided with a regulating door, through which the dirt and surplus water is drained off. The arrangement of the apparatus is shown in Fig. 17.

By regulating the upward velocity of the water so that the volume of water added to that in the trough is equal to, or slightly greater than, that of the shale extracted, no perturbation of the flow of water in the trough is caused and no suction is exerted on the upper layer of the washing bed, so that no coal is drawn through the opening in the trough. A slightly greater supply of water will ensure that a proportion of the shale passes over the opening with the coal, while if the supply is reduced below the normal, all the shale, together with a small proportion of coal, will be withdrawn.

A pipe is fitted to the box through which a supply of water is admitted. The amount of water is regulated by a cock so that the water level in the washing trough remains constant, the flow of water passing under the central diaphragm and up through the slot in the trough through which the shale is extracted. In this way the impurities can be entirely removed, but in practice it is found that it is advisable to carry out the work in two or more stages. The first set of Rheolaveurs are so adjusted that a small proportion of coal is allowed to be carried out with the dirt, and this material is then passed through a second trough, where a further washing takes place. In this second trough the last two or three Rheolaveurs are also adjusted to pass a small proportion of coal, and the material from these outlets is passed into a third trough, where the flow of water is so adjusted that shale only is allowed to pass through the outlets, with a small proportion of shale also passing over the end with the coal. This material is elevated and fed back into the first trough with the raw coal. The clean coal delivered from the washer is carried forward by a conveyor which is constructed of wire mesh or similar material, so that the surplus water can be drained off.

If a coal of mixed size is being treated the small coal may contain a proportion of dirt. In that case the meshes of the conveyor are made large enough to pass the smaller sizes of coal, which are then elevated and given a further separate treatment. The larger coal is passed forward to screens for separation into various sizes.

This type of washer is particularly suitable for the washing of fine material and is unaffected by variations in the quality of the coal delivered to it.

This result is obtained by arranging a sufficient number of Rheolaveurs to ensure that even under the worst conditions the coal leaving the end of the trough is free from shale. If a coal containing a smaller proportion of impurities is supplied to the washer, the products evacuated by the last two or three Rheolaveurs will be clean, whereas in the worst case possibly the last Rheolaveur only will evacuate dirt-free coal. The products

from the first set of Rheolaveurs, which have been
set normally to withdraw material containing a

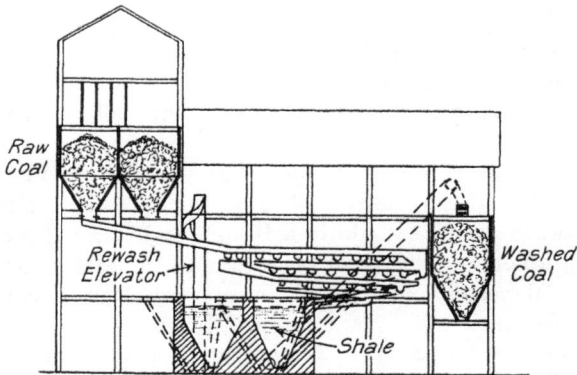

General Arrangement of Rheolaveur Washery

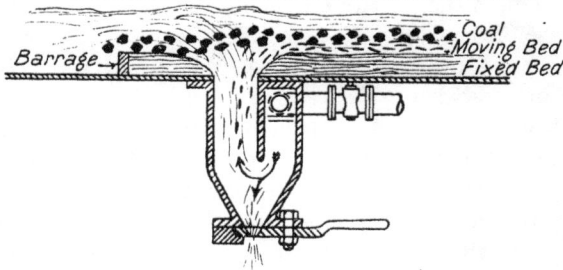

Detail of Rheolaveur

FIG. 17.—Rheolaveur Washer.

small proportion of coal, will remain normal, and
this material will be retreated in the usual way.

The description given above deals particularly

with the Rheolaveur Washer as used for dealing with fine coal.

When grain coal is to be washed a modified pattern is used in order to reduce the amount of water required. In this case a sealed type of Rheolaveur is used, the shale evacuated by the Rheolaveur being deposited into the closed boot of an elevator.

The principle of washing of grain coal is similar to that of the washing of fines, that is to say, the classification is made in a trough under the effect of a horizontal current of water and the extraction of deposited matter by means of Rheolaveur apparatus. The plant in itself is a little different, however. It consists, in general, of a single washing trough fitted with two or three Rheolaveurs. These Rheos themselves are designed differently, although the principle remains the same as in the case of the fines. The classification of the particles above $\frac{1}{4}$ in. or $\frac{5}{16}$ in. is effected in a relatively short distance. This allows of reducing considerably the length of the trough. On the other hand, the phenomena of surface tension, viscous resistance, etc., which play a notable part in the treatment of very fine coal, have no apparent action when washing grain coal, and it is therefore not necessary to split up the classification into successive phases (as in the case of washing fines) when treating the larger sizes.

The same type of Rheo as applied in the case of fines, could be used, but the quantity of water passing through the evacuation orifice would be too considerable by reason of the diameter of this orifice being large enough to pass this size of coal without obstruction. In this case use has been made of an apparatus called " Niveau Plein," or sealed type Rheolaveur, in which the evacuation orifice of the Rheo is fitted direct on to a sealed elevator.

In this manner the shale is dealt with without loss of water.

A washer of this type, having a trough 16 ft. long by 1 ft. 8 in. wide, can deal with 100 tons per hour of raw material, with a consumption of about 25 H.P. for the pumps, elevators and draining conveyor. The amount of water circulated would be about 18,000 gallons, but the greater part of this would be used over again and only a small quantity of make-up water would be required.

The Rheolaveur Washer is made in Great Britain by The Butterley Co., Ltd.

The " Nota Nos " Washer, made by Messrs. Head, Wrightson & Co., Ltd., of Stockton-on-Tees, is largely used for coal washing and for the separation of the combustible material from the pan ashes of retort settings. It consists of a long water-tight trough, built up of steel plates, angles and tees, and arranged at a slight angle. The trough is carried on chilled cast-iron rollers, supported on channel steel stools, the rollers being held in position by steel spindles coupled together by springs. The driving mechanism consists of the " Marcus " propulsion gear, previously described in connection with the Marcus Screening Conveyor, connected to the trough by a cast steel eccentric rod, so that the trough in its forward motion is

3*

gradually accelerated, and gradually retarded during the return stroke.

The arrangement of the Nota Nos Washer is shown in Fig. 18. The coal or pan ash is delivered from a hopper arranged over the screen at the driving end on to a feed tray, along which it is conveyed into the washer box where it meets a flow of water. The coal is separated from the dirt, and is carried down the trough over an adjustable weir on to a lower tray fitted with perforated plates or screens constructed of wedge-shaped rods or wires, according to the size of material which is being handled. The water and any fine coal are led off to a settling tank, while the clean coal is conveyed forward to the end of the lower tray or deck, whence it is conveyed to the screens or storage bins. The dirt and shale separated from the coal are conveyed against the flow of the water to the upper end of the main washing trough and are then delivered into a conveyor or removed in any other convenient manner. The water is supplied through sprays arranged over the upper end of the trough. This water is used repeatedly, and about 16 gallons of fresh water only are required per ton of coal washed. The fine coal deposited in the settling tank, if sufficiently free from ash, may be mixed with the washed small coal, or may be burnt independently under the boilers.

The power consumption is approximately 0·8 H.P. per ton of coal washed per hour.

Rotary and Reciprocating Screens may be used for the washing of gravel and the coarser sands,

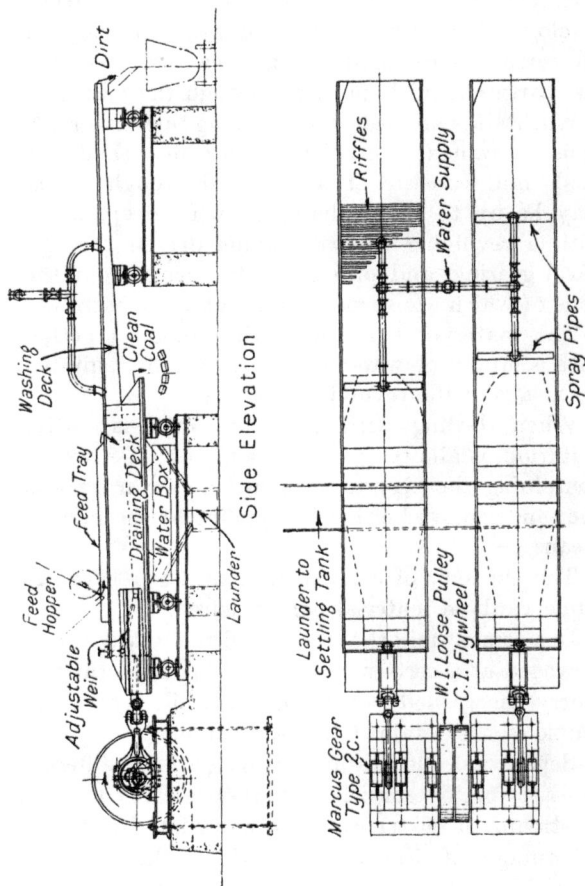

FIG. 18.—Nota Nos Washer.

but are not suitable for dealing with very fine materials owing to the tendency of the fine screens to clog. A plentiful supply of water playing on the screens is required in either case to wash the fine particles and impurities through the meshes.

Wash Mills are used for screening fine materials, such as cement slurry and very fine sand. A wash mill consists of an annular trough which may be up to 15 ft. diam. by 6 ft. deep, fitted with a revolving central spindle driven through bevel gearing, and provided with arms from the ends of which are suspended agitators or harrows.

The material is run into the trough together with a supply of water and is thoroughly agitated by means of the revolving harrows.

When dealing with sand the material, after agitation, is allowed to settle, and the wash water, containing the clay in suspension, is run off and the sand removed either by hand or by mechanical means.

The capacity of a large wash mill is 100 to 120 cubic yards of material per 24 hours.

In a cement mill the trough is fitted with gratings arranged at a certain level, through which the fine slurry is washed continuously, while any large particles are retained in the trough.

Separation and grading of materials of different sizes or specific gravities can also be carried out in a stream of water without agitation by taking advantage of the rate at which the different particles will settle under the action of gravity when they are suspended in a liquid. The rate of settlement depends solely on the size and specific gravity of the particles and the density and viscosity of the liquid, and it is therefore possible so to regulate the flow of the water that the different sizes or classes of material are deposited separately.

The rate at which settlement takes place is determined as follows :—

Weight causing acceleration $= \frac{\pi}{6}D^3w(S - s)$ (1)

Where D = diameter of body in inches,
S = specific gravity of body,
s = ,,   ,,   ,, liquid,
w = weight of unit volume of water.

Resistance to motion $= KAws\frac{V^2}{2g}$ . . . (2)

Where A = projected area of body,
V = velocity in inches per second,
K = a coefficient varying with the form of the body but having an average value of 0·5 as determined by various experimenters.

Substituting in (2) above we have for an approximately spherical body

Resistance $= 0·5\left(\frac{\pi D^2}{4}\right)ws\frac{V^2}{2g}$ . . . . . (3)

Equating (1) and (3)

$$0·5\left(\frac{\pi D^2}{4}\right)ws,\frac{V^2}{2g} = \frac{\pi}{6}D^3w(S - s),$$

$$\frac{V^2}{1030} = D\left(\frac{S - s}{s}\right),$$

$$\text{or } V = 32·1\sqrt{D\left(\frac{S - s}{s}\right)} \text{ inches per sec.},$$

$$\text{or } V_1 = 2·67\sqrt{D\left(\frac{S - s}{s}\right)} \text{ feet per sec.}$$

In metric units :

$$Vm = 5 \cdot 11 \sqrt{Dm\left(\frac{S-s}{s}\right)}.$$

Rittinger gives the velocity in metric units as

$$Vm = 3 \cdot 91 \sqrt{Lm\left(\frac{S-s}{s}\right)}$$

and the ultimate velocity of fall in metres per second for bodies of various shapes by the following :—

$$Vm = 3 \cdot 2 \ C \quad \text{for rounded bodies,}$$
$$Vm = 2 \cdot 25 \ C \quad \text{for flattened bodies,}$$
$$Vm = 2 \cdot 65 \ C \quad \text{for elongated bodies,}$$
$$Vm = 2 \cdot 85 \ C \quad \text{for average bodies,}$$

where $C = \sqrt{d_m\left(\frac{S-s}{s}\right)}$

d_m being the diameter of the body in metres and S and s being, as before, the specific gravities of the solid and liquid respectively.

The viscosity of the liquid used also affects the result, but the figures above refer to water, which is the liquid most commonly used.

Various machines have been designed to employ this property of settlement at known rates. Several of these machines consist of conical vessels, known as Spitzkasten and Spitzlutte, into which the material is fed at one side from the top, while a stream of water is admitted at a carefully regulated rate from the bottom. The fine particles overflow through an opening near the top with the water, while the heavier material sinks to the bottom and is withdrawn at intervals through a second opening at the apex of the cone. At the narrowest part of the vessel the rate of flow is sufficient to agitate the material thoroughly and to separate the different sized particles, while the velocity of the water gradually decreases as the diameter of the vessel increases.

A series of these vessels may be used for classifying materials into various grades or sizes, the overflow from one vessel passing forward into a similar washer of larger diameter. The rate of flow of the water is regulated separately in each vessel to give the desired separation.

The Allen Cone Washer is of a similar type, but is provided with an automatic valve, governed by a float, which automatically regulates the discharge of the coarser material which settles to the bottom of the tank. The valve opens as soon as a certain amount of material has collected, and closes again when all the material has passed through and clear water commences to flow.

This type of washer can be used either for classifying a material into its different sizes, or for separating different materials by taking advantage of their varying specific gravities.

The Dorr Thickener operates on a somewhat similar principle, and is used for the recovery of the solid materials from the overflow and drainage from coal washeries, so enabling the same water to be used over and over again without the use of settling tanks. It has the advantage of taking up much less space than ponds or tanks, and the power required for its operation is only small.

The Dorr Thickener is in extensive use in America, but there are only a few examples in

Great Britain. It consists of a circular tank with the bottom sloping towards the centre where the sludge outlet is situated. The water from the washery is run in at the top near the centre and the clear water overflows into an annular channel arranged round the circumference.

From a vertical arm in the centre a number of radial arms are suspended, and by means of these arms the sludge is continually stirred up and gradually settles towards the centre of the tank, from which it is drawn off through the sludge valve.

The centre shaft with its arms can be raised or lowered so as to avoid the possibility of the arms being jammed by an accumulation of sludge. It is revolved by means of a motor and suitable gearing.

A tank 50 ft. in diameter by 8 ft. deep will deal with 6000 gallons of water per hour, containing 3000 to 6000 lb. of solids, while a tank 200 ft. diameter will handle 5 to 6 million gallons per day, containing 3000 to 4000 tons of solid material. The power required for operating the larger machine, apart from any pumping necessary, is only about 3 B.H.P.

The solid material which is extracted contains a considerable percentage of combustible matter, which can be used as fuel after it is sufficiently dried.

The Dorr Thickener is also used as a Classifier, as by suitably proportioning the tank the water can be run off while it still contains more or less of the finer particles. It is used for this purpose in cement works.

The moisture in the sludge may vary from 25 to 65 per cent., according to the nature of the material and the absence or presence of clay.

Messrs. Simon Carves, Ltd., employ large conical settling tanks for clarifying the water from their coal washers. The tank is arranged at a suitable level so that the clean water supply to the washery can flow by gravity from an outlet near the top of the tank to the washery boxes. The water from the washery is pumped back through a pipe having its outlet in the centre of the tank and at a point slightly higher than the water level. The outlet of this pipe is surrounded by a steel cylinder or curtain, which has its lower edge perforated in order to guide the suspended material downwards without contaminating the upper layer of clean water in the tank.

The velocity of the incoming water is gradually reduced, thus allowing the fine suspended material to settle down to the bottom of the tank. An overflow pipe is provided so that the water level is kept constant, and a further pipe is provided for bringing in fresh water to replace the amount which is carried away with the washed coal.

An outlet is arranged in the bottom of the tank, provided with a cock, so that the sludge can be passed out by gravity and fed back to the washery to be mixed with the fine coal and re-washed. Only one pump is required for keeping the water in circulation.

When Settling Tanks or Ponds are used, the area must be large enough, in proportion to the volume of water passing through the plant, to allow the solids ample time to settle out. Unless the velocity of the water is sufficiently reduced the solids will not be deposited.

The drawbacks to the use of settling tanks, particularly when large quantities of materials are to be dealt with, is the space occupied and the fact that the mud has from time to time to be dug out and removed by hand.

The Dorr Classifier consists of an inclined trough open at the upper end but closed at the bottom. The feed enters near the centre and the water overflows at the closed end, carrying with it the lighter particles of material, while the heavier material settles to the bottom and is gradually conveyed up the incline by mechanically-operated rakes. The upper part of the bottom plate is above the level of the water in the trough, and allows the surplus moisture to drain back before the material is discharged over the open end.

These Classifiers vary in size from 15 in. wide by 10 ft. long to 6 to 8 ft. wide by 26 ft. 8 in. long, and are mainly used for classifying fine ores, removing slimes from sands, and for de-watering material down to a moisture content of 15 to 30 per cent.

The Dorr Bowl Classifier is of similar type, but carries out the separation in two stages. A large shallow bowl is fixed over the deep end of the trough and the feed is introduced into the centre of this bowl. The larger particles of material which settle out in the bowl are discharged into the inclined trough through a small opening by means of a revolving rake, while the surplus water and fine slimes overflow round the edge of the bowl into a circular duct or launder. An auxiliary supply of water is admitted from a spray pipe near the head of the inclined trough, in which a further washing and separation take place, the clean, oversize material being discharged, as before, by the mechanical rakes over the open head of the incline.

Pan Ash Washers

With modern methods of carbonisation there is a decrease in the amount of coke available for sale, and it is therefore important to conserve all the coke available, even if it is not in a condition suitable for sale, as it may be used in the boiler furnaces of the gas works. An appreciable quantity of coke is present in the ash recovered from the gas producers in the retort settings, and various means are employed for extracting this coke.

The machines used for this purpose are of three general types, the most usual being one employing water and taking advantage of the difference in the specific gravities of the coke and the ash and clinker. Magnetic methods have also been used, and a process employing a slurry of adjustable specific gravity.

The Robinson Pan Ash Washer consists of a conical vessel in which a vertical shaft rotates. This shaft is driven by gearing and carries a number of stirring arms. In a plant of this type recently installed the vessel is 5 ft. diam. at the top and 5 ft. deep, tapering to 12 in. diam. at the bottom.

A 4 in. diam. pipe perforated with $\frac{9}{16}$ in. holes is arranged just above the outlet, water being supplied to this ring at a pressure of 11 lb. per sq. in. The pan ash is fed into the conical vessel by means of an elevator and the ash and water are continually stirred by means of the revolving arms. The ascending stream of water separates out the coke and breeze, which is washed over a second outlet arranged near the top edge of the vessel and down a shoot into a suitable bin. The upper part of the shoot has a perforated bottom plate having holes $\frac{1}{4}$ in. diam., in order to drain out the water and very fine material, which is led to a separate tank. A filter prevents the greater part of the sludge from getting into the main water tank. The outlet at the bottom of the conical vessel is provided with a water-lock chamber about 12 in. diam. by 12 in. deep, fitted with doors at the top and bottom, so that the clinker which settles to the bottom of the cone can be removed without the loss of any considerable quantity of water.

A test gave the following results :—

Pan ashes treated   .   . $48\frac{1}{2}$ tons.
Coke and breeze recovered 21·2 ,,   = 43·5%
Clinker and ashes   .   . 25·8 ,,   = 52·1%
Silt   .   .   .   . 2·2 ,,   = 4·1%

The current consumption was approximately 5 units per hour, including the power required for elevator, washer and pump. The amount of pan ashes made per bed of eight retorts is approximately 1 ton per day, and over 1300 tons of coke and breeze were recovered during a period of about six months. The recovered material contains about 23 per cent. of moisture when removed from the plant, but after drying for three days this is reduced to 14 to 15 per cent., and the material is then quite suitable for using under boilers.

The Nota Nos Washer, previously described, has been used very successfully in a number of gas works for separating the ash and combustible material in pan breeze.

The Columbus Pan Ash Separator is now used in many gas works for separating the combustible material from the ash. The Columbus Separator consists essentially of a rotary screen for eliminating dust and large lumps of slag, and the separator proper, which utilises a liquid having a specific gravity of 25° to 30° Beaumé, in which the actual separation of the coke and slag takes place. The furnace ashes are fed into the rotary screen by means of an elevator, or in any other convenient manner. The first section of the screen is covered with fine-mesh sheets and serves to separate out the fine dust, the larger material passing forward into the second portion of the drum, which is provided with screen bars or plates having openings of a size suitable for passing all the material except the largest lumps, which are generally pure slag having no calorific value.

The material which has passed through the openings in this second section falls into the lower end of the separator, which consists of an inclined chamber built of steel plates. This chamber has a central division plate dividing it into two inclined compartments one above the other, in each of which there is a worm conveyor. The separator tank is half full of the liquid, usually water mixed with clay, chalk or any other available substance which will increase the density to the required amount.

As the ashes fall from the screen into this liquid, the heavy slag falls to the bottom of the tank, from which it is removed by the lower worm,

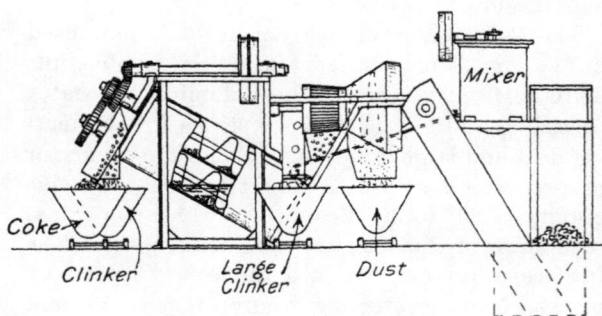

FIG. 19.—Columbus Pan Ash Separator.

which extends to the bottom of the tank. The lighter coke floats on the surface of the liquid and is picked up by the upper worm, which projects into the liquid but does not reach the bottom of the tank. As the coke is above the surface of the liquid during the greater part of its travel up the worm conveyor, the surplus moisture is able to drain back into the tank. The two conveyors deliver the separated materials into separate shoots, from which they can be fed into wagons or hoppers as required.

Columbus Separators are made in various sizes, having capacities from 2 to 8 or 10 cubic yards of material per hour. The smaller size requires about 1 to $1\frac{1}{2}$ H.P. and uses about 20 gallons of water per hour, while the largest size requires 3 to 4 H.P. and uses about 50 gallons per hour.

The arrangement of a small Columbus Separator is shown in Fig. 19.

CHAPTER V

As mentioned previously, it is frequently necessary in the case of mineral-bearing sands and ores to concentrate the material before treatment for the extraction of the valuable mineral, in order to remove as much as possible of the adventitious matter. The non-mineral bearing portion is generally of lower specific gravity, and can conveniently be treated with water to effect its separation.

The simplest form of concentrator consists of a rectangular or circular box or tank, with an inclined bottom. The fine ore and water are fed in at the highest point and continuously agitated by hand or by mechanical means. The heaviest particles settle more rapidly, and are therefore deposited at the upper end of the slope, while the lighter quartz particles are carried to the lower part of the apparatus. The surplus water overflows at the lower end in the case of a rectangular tank, or round the outer circumference in the case of the circular type. This form of apparatus is generally known as a " Buddle."

The deposited material must be removed from time to time by digging, and for satisfactory results the amount of water admitted and the rate of flow of the material must be determined by experience for each class of ore.

Sluices are similar to rectangular buddles, but are made narrower, being usually from 1 to 2 ft. wide only, and from 12 to 15 ft. long, the angle at which they are set being governed by the nature of the material.

Table Concentrators are also of a somewhat similar type, but are about 4 or 5 ft. wide and about 12 ft. long, and are generally agitated mechanically while the material mixed with water is flowing over the surface of the table. The surface is covered with blanket or other textile material, or provided with a series of ridges or riffles against which the heavier particles are collected, while the lighter material is washed away. In some types the tables are given a longitudinal motion, while in other cases the movement is transverse.

In the latter pattern the material is spread over the table by the reciprocating movement and the various grades of ore are carried by the motion into different channels according to their specific gravity.

These tables are largely used for concentrating gold and tin ores after they have been finely crushed by stamping.

The Wilfley Ore Concentrator is one of the best-known machines of this type. It consists of a flat inclined table, about 16 ft. long and 7 ft. wide at the driving end, tapering off to about 3 ft. at the other end, on which are a series of riffles of varying length.

The pulp or slime is fed on to the table near the driving end and flows down the inclined table towards the tailings discharge. The table has a differential motion, rapid at the outer end of the stroke and slower at the inner end. This motion causes the particles of ore to settle in layers, leaving the silica at the top exposed to the action of the washing water, while the whole mass of material moves slowly up the inclined table towards the concentration box. As the pulp moves up the table the upper layer of sand is carried over the riffles by the water, while the heavier concentrates settle in the grooves and a fresh layer of sand is brought to the surface. The concentrates gradually move along the grooves to the end of the riffles, where they emerge on to a plain part of the table. Here they meet a stream of fresh water, which carries the remaining sand into the tailings discharge and also washes the minerals down the table in the order of their specific gravity. The longitudinal motion of the table moves them at the same time towards the concentration box, so that the various concentrates form distinct zones and can be discharged into the separate compartments of the box.

The vibratory motion is produced by toggles actuated by an eccentric on the driving shaft, working against the pressure of a spring which forces the table back on the return stroke. The table is covered with linoleum and is made adjustable, so that the angle of inclination can be regulated to obtain the best results, by means of wedges.

The speed of the driving shaft is about 240 r.p.m., and the stroke varies from $\frac{3}{8}$ to $\frac{3}{4}$ in., the output being up to 2 tons per hour, according to the nature of the material.

The Buss Patent Concentrating Table, made by Messrs. The Coal and Ore Dressing Appliances, Ltd., is somewhat similar to the Wilfley Concentrator, but the inclined table is carried on a number of ash strips or legs in a similar manner to the Zimmer Conveyor. These legs are secured to a steel bottom frame, which can be tilted as required in order to vary the inclination of the table both transversely and longitudinally. The table is driven by means of an adjustable eccentric, so that the stroke can be rapidly altered to any length required. The use of the spring legs results in the table having a vertical as well as a horizontal movement and the amount of upward throw can be varied, if necessary, while the machine is running, by altering the length of the eccentric rod, which controls the inclination of the legs.

The surface of the table is covered with linoleum and may be fitted with riffles, or left plain, as may be most suitable for the ore under treatment.

The water-feed trough extends the whole length of the table and is divided into sections, controlled separately, so that the amount of water flowing on to different parts of the surface may be varied. The receiving launders for the products and tailings are also divided into sections by adjustable trays, and vibrate with the table in order to prevent clogging.

The speed at which the machine is run varies from 250 to 290 r.p.m., according to the class of ore which is being treated. The capacity varies from 12 to 36 tons per day, according to the material, and the power absorbed from $\frac{1}{3}$ to $\frac{3}{4}$ B.H.P. This table is particularly suitable for dealing with materials from 30 to 40 mesh in size.

The Frue Vanner, made by Messrs. Fraser and

Chalmers, Engineering Works, Ltd., of Erith, has a very wide application for the fine concentrating of materials. It does not require the material to be sized, but separates the ore only into two grades, the concentrates and the tailings, and is therefore suitable for handling those ores only where one mineral is to be concentrated.

It consists of an endless rubber belt, 4 or 6 ft. wide by 12 ft. long, with rubber flanges, which travels up a slight incline and is carried by drums. The belt dips into a water tank, in which the concentrates are collected, and is given a gentle shaking movement from a shaft which runs along one side of the frame. The fine ore is fed on to the belt by means of a spreader at a point about 3 ft. from the head, together with a stream of water. As the belt travels, the shaking motion separates out the sand and mineral, the latter travelling along with the belt while the lighter sand is washed down the incline and discharged at the lower end. The inclination of the belt can be varied from $\frac{1}{4}$ to $\frac{1}{2}$ in. per foot. Increasing the inclination of the belt makes the bed of pulp on the table thinner and tends to give cleaner concentrates, while decreasing the slope has the opposite effect. Increase of speed also cleans the concentrates and thins the bed of pulp.

At a normal rate of working a Frue Vanner 4 ft. wide will handle about 1000 lb. of concentrates per 24 hours.

The amount of water used varies from $2\frac{1}{2}$ to $4\frac{1}{2}$ gallons per minute.

The Embrey Concentrator is of similar type,

but has a longitudinal shaking motion instead of a transverse movement.

The Luhrig Vanner consists of an endless rubber belt supported by small intermediate rollers and two larger rollers at the ends. The rollers are fixed on a frame which is suspended at the ends from fixed supports. These supports are adjustable, so allowing the inclination of the table to be adjusted. The frame is set level longitudinally, but slopes from side to side, and is given 150 to 200 strokes per minute by means of a cam on the driving shaft working against a spring at the other end of the frame. The length of stroke may vary from $\frac{1}{6}$ to $1\frac{1}{2}$ in.

The belt travels at a speed of 8 to 10 ft. per minute, and the ore is fed on to it through a feeder at the top corner of the belt. The feeder has hollow brushes which come down on to the surface of the belt, and the pulp passing through these brushes is spread closely on the belt and any air bubbles are broken up. A spray pipe is fixed diagonally across the belt. At the front of the machine is a receiving launder divided into separate compartments by movable slides, into which the various products flowing down the belt are discharged and from which they are led by separate pipes. The middlings can be led back for retreatment on the same machine, or collected for separate treatment if preferable.

The power required for driving the machine is about $\frac{1}{3}$ H.P.

A Luhrig Concentrator 12 ft. by 3 ft. 6 in. travelling about 10 ft. per minute, and struck

160 to 180 blows per minute, has an output of 2 to 8 tons per 24 hours, according to the nature of the material.

The Buss " Slimer " is intended for use with extremely fine materials which cannot be handled successfully on a concentrating table, and particularly for materials below 80 mesh in size. It is somewhat similar to the Frue Vanner, described above, but embodies a number of novel features, and employs a special canvas composition belt which has a rough surface instead of the rubber belt used in the older pattern of machine. The endless belt is 5 ft. 6 in. wide and passes over end rollers mounted on a steel frame which is carried on wooden spring legs, bolted at their lower ends to a rigid bottom frame. The bottom frame can be tilted longitudinally or transversely by means of screws.

The upper frame carrying the belt is held up to a cam wheel by means of springs and tension rods, the lower ends of which are attached to the bottom frame. The cam wheel pushes the frame against the pull of the rods and on release the frame is drawn back rapidly against a buffer fixed on the driving standard. The buffer is made adjustable in order to regulate the amount of " bump," and the pressure of the frame is also adjusted by altering the length of the tension rods.

The belt is carried on a wooden bed, provided with diagonal grooves which are supplied with water from a longitudinal groove running along the upper edge of the bed. This results in a minimum amount of friction and wear, as the belt is carried on a film of water.

A feed-box and water supply are arranged at the upper edge of the bed, and a receiving launder at the lower edge, divided into compartments, as usual, for the concentrates, middlings and tailings.

The Buss Slimer is generally made with a length of 15 ft. and an overall width of 7 ft. 6 in. The capacity is from 6 to 12 tons per 24 hours, according to the class of material under treatment, and the power absorbed $\frac{1}{3}$ to $\frac{1}{2}$ H.P.

This machine is made by Messrs. The Coal and Ore Dressing Appliances, Ltd.

The " H.H." Universal Concentrator is used for separating and cleaning concentrates, sands and mineral ores, and has also been successfully applied to the cleaning of fine coal up to $\frac{3}{8}$ in. in size. This concentrator embodies the Overstrom unbalanced pulley agitating gear, which has been previously described, for giving the necessary transverse motion to the table. The " H.H." Concentrator is supplied by Messrs. Hugh Wood & Co., Ltd., of Newcastle-on-Tyne.

The Concentrator consists of a shaking frame, supported on flexible legs constructed of strips of hickory or ash rendered waterproof and wrapped with tape. The supporting castings, both on the rigid floor and the table, are fixed so that the edges of the strips point radially to a common centre, and at one end of the table a diagonal brace, also of hickory or ash, is connected at one of its ends to the table and at the other to the fixed support, so that as the table reciprocates its motion is slightly curvilinear.

At the head end of the Concentrator a heavy

yoke is rigidly fitted to which is clamped the fixed hollow shaft carrying the unbalanced pulley, which is located by means of collars on the shaft and is lubricated by a grease cup screwed into the end of the shaft.

Pressing against the crosspiece of the yoke are two springs which assist in accelerating the forward and retarding the backward strokes, the other ends of the springs pressing against two hinged posts, by means of which the pressure of the springs and the stroke of the table can be adjusted.

A bumping post is fixed rigidly to the frame and provided with a cushion or pad, against which the table is brought to rest on its forward stroke. This pad is of special construction to ensure cool and noiseless working, and means are provided for regulating the length of the stroke while the machine is running.

By screwing up the tension rods which adjust the pressure of the hinged posts on the springs the table can be brought to rest while the pulley is still running. The table can be operated with one spring only in use if necessary, so as to avoid a stoppage should one of the springs break.

The table top, which is lined with linoleum and fitted with riffles, is hinged to the frame at one edge, and provided with a series of wedges operated by a screw at the other edge, so that the inclination of the table can be varied as required.

The concentrates fall into a launder, which moves with the table, and they leave the launder through two pipes.

The feed- and water-box is fixed along the top

edge of the table, and arranged so that the feed can be stopped off at any point according to the amount of water in the pulp.

The riffles are curved and extend nearly the whole length of the table. The back flow of the water arrests the forward motion of the sand or coal, while the heavier material is carried forward beyond the ends of the riffles and held well up on the table and away from the sand or coal.

The power required depends on the amount of material fed on to the table, about $\frac{1}{4}$ H.P. being required for light loads and $\frac{3}{4}$ H.P. for heavy loads. The standard table is 6 ft. wide by 9 ft. long, but for coal washing a 7 ft. wide table is supplied. The capacity of the standard machine varies from 15 to 180 tons per 24 hours, according to the size of the material, the concentration, the difference in specific gravity between the concentrates and the sands or other refuse material and the condition of the pulp. The average output may be taken as 60 tons per day.

The " H.H." Concentrator is very simple in construction and operation, and will handle material varying in size from $\frac{3}{4}$ in. to slimes, according to the arrangement of the riffles.

The following results were obtained from tests on coal 0 to $\frac{3}{8}$ in. in size containing 30 per cent. of ash.

Sample from coal end of concentrator :—

Clean coal	96·9 per cent.
Free dirt (over 1400° S.G.) .	3·1 ,,
Ash in clean coal . . .	6·0 ,,
Ash in free dirt . . .	44·0 ,,

4

Sample from stone end of concentrator :—

Coal	1·3 per cent.
Free dirt (over 1400° S.G.) .	98·7 ,,
Ash in coal	28·5 ,,
Ash in free dirt . . .	79·33 ,,

The capacity of the concentrator, when working on small coal, is from 5 to 7 tons per hour, varying according to the size of the material and the proportion of refuse with the coal.

The arrangement of the " H.H." Concentrator is shown in Fig. 20.

Several other machines are made on similar lines, embodying modifications in the method of drive, but employing the general principles described above.

Jigs consist essentially of trays of perforated plate or wire mesh on which the material is placed in a trough of water, the water being given a pulsating motion by means of a piston placed in a chamber or cylinder connected to the trough. The principle on which they operate is the tendency of particles of approximately similar size, but of different specific gravities, to arrange themselves in layers according to their gravities when given motion by means of a pulsating column of water, the lightest material rising to the top and the heaviest particles falling to the bottom.

In actual practice a Jig usually consists of a tank constructed of steel or wood, with a partition extending across the upper part to about half the depth of the tank. On one side of the central partition is fixed the horizontal screen on to which the ore is fed, while in the other portion of the tank one or more plungers or pistons are arranged, working vertically and operated by cranks or eccentrics from an overhead shaft.

The movement of the plungers causes a regular pulsation of water through the screen and agitates

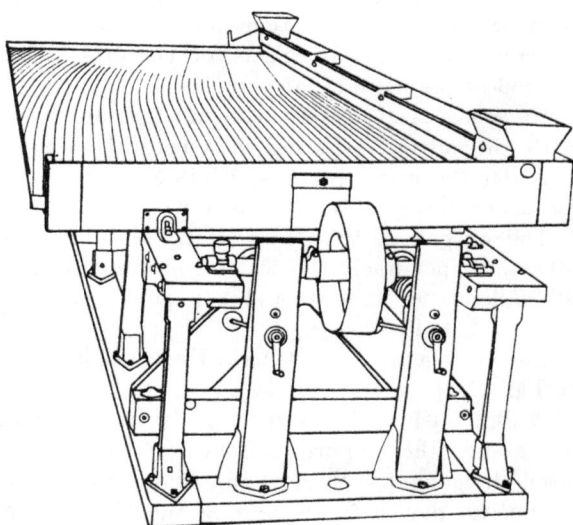

Fig. 20.—" H.H." Concentrator.

the bed of ore resting thereon, this agitation allowing the heavier particles of ore to settle down through the lighter quartz. The ore particles are either drawn down through the screen or are discharged through an opening at a suitable level, while the lighter material is carried forward by the flow of the water and is discharged over the end of the Jig through an opening at a higher level than that arranged for the discharge of the concentrates.

This arrangement is generally known as the Hartz Jig.

The Baum Washer, used largely for the washing of coal, and described previously, operates in this manner, but the pulsations of the water are effected by means of blasts of compressed air instead of the more usual plungers.

In some cases 2-compartment Jigs are used, with the piston operating in a central chamber between the compartments, while 3- and 4-compartment Jigs are also in general use.

The output of this type of Jig is from 5 to 10 cu. ft. per hour, the water consumption 6 to 12 gallons per minute and the power required about $\frac{1}{2}$ B.H.P.

The arrangement of a typical Hartz Jig is shown in Fig. 21.

A Jig Washer, designed specially for removing stone and other impurities from coal, has recently been designed by M. de Caux and embodies an automatic device for regulating the thickness of the filtering bed. This machine is known as the Autolaveur and is of the Hartz type, the pulsation of the water being produced by plungers in the usual way. The special automatic feature, which results in a considerable increase in the percentage of clean coal, consists of the fitting to the screen side of the apparatus of a number of automatic valves controlling the outlets for the stone.

The automatic valve consists of a cylindrical

aluminium tube containing a piston, the position
of which can be regulated by a screw.   This tube
floats in the water and slides in an outer tube
attached to a casting bolted to the screen, the
casting having openings through which the stones
fall into the discharge pocket.   At each pulsation
of the water the aluminium tube rises, carrying
with it the piston, which opens the stone discharge
outlet to a greater or less extent, according to the

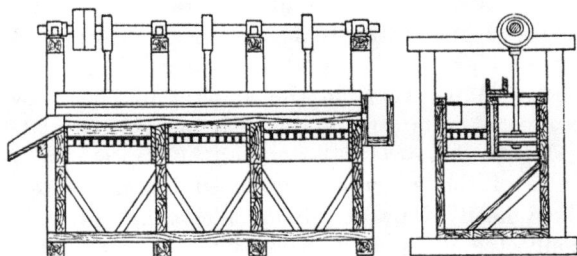

Fig. 21.—Three-Compartment Hartz Jig.

position of the piston in the tube and the resistance offered by the bed of stone lying on the
screen.   The rate of discharge of the stone thus
varies according to the height to which the piston
is lifted.   With a thin bed of stone, the resistance
offered to the pulsation of the water is small and
the piston is only lifted to a comparatively small
extent.   As the thickness of the layer of stone
increases, the resistance offered to the ascending
current of water is increased and the piston is
lifted to a greater height, thus increasing the
discharge opening and allowing a greater quantity
of stone to pass away.

This action is quite automatic and regulates the thickness of the filtering bed to the predetermined amount. Jigs fitted with the Autolaveur apparatus are therefore entirely unaffected by changes in the rate of feed or of the amount of free stone present in the coal.

The coal is screened before entering the Jigs, in order to remove the larger coal which is to be hand-picked, and to ensure that each Jig receives coal of a more or less uniform size. Dust is also removed before the coal is washed. Tests made over an extended period showed an increase of about 6½ per cent. in the amount of clean coal recovered from a given quantity of raw material.

In the Richards Pulsator Jig the upward pulsations of the water are produced by means of a fixed head of water, giving a constant pressure, connected at rapid intervals by means of a rotating valve to the tank containing the screen on which the bed of ore rests.

This Jig occupies less space and requires a smaller quantity of water than Jigs of the Hartz type.

A single-compartment Richards Jig, 12 in. by 12 in., has a capacity of 200 to 300 tons per 24 hours, with a consumption of 150 to 225 gallons of water per minute, and requires about 1·3 B.H.P.

A series of these Pulsators fixed to the bottom of a long trough or launder about 10 in. wide by 10 in. deep enables the ore to be classified into several sizes, each Pulsator being arranged to give a product of a separate definite weight, the heaviest particles being withdrawn in the first and the lightest in the last settling pocket.

The Richards Jig is made by Messrs. Fraser and Chalmers, Engineering Works, Ltd.

The Hancock Jig consists of a rectangular wooden tank in which is placed a sieve to which is given a special motion. At every stroke the sieve is lifted vertically a distance of $\frac{1}{4}$ to $\frac{3}{8}$ in. and at the same time it is given an oblique forward movement of about the same amount. The tank is kept nearly full of water and a small supply of water is usually allowed to flow in, though, if necessary, only sufficient additional water may be used to ensure the removal of the concentrates and tailings into their respective compartments, as the forward motion of the material is effected solely by the movement of the sieve.

This type of Jig is therefore particularly suitable for use in districts where water is scarce.

Several other types of Jigs are in common use, differing only in the arrangement of the tank and sieve and on the method of producing and applying the pulsations.

The output of Jigs is affected by the nature of the material, the ratios of the specific gravities of the ore and rock and of the sizes of largest and smallest particles, and is also greatly influenced by amplitude and frequency of the oscillations of the water and the manner in which they are produced and applied to the bed of ore.

Froth Flotation Process

The concentration of ores, or the separation by flotation of the minerals from the quartz or other

rock with which they are mixed, depends on the
fact that many minerals, if finely ground, will
float on the surface of a suitable liquid, even
though they are of greater specific gravity than
the liquid, due to the difference of surface tension,
and by suitable arrangements it is possible to
ensure that almost the whole of the mineral or
ore can be kept in suspension, while the heavier
quartz sinks to the bottom. The valuable ore
can then be removed, leaving behind the rock
with only a small percentage of the mineral.

Various processes are in use employing oils,
water and grease as the separating media, but the
process in most general use is that known as the
Froth Flotation process.

In this system the area of the liquid in contact
with the material is greatly increased by churning
the liquid into froth, or by aerating it in any
convenient way. The liquid used consists of
water with a small quantity of an aerator, which
may be soap, turpentine, cresylic acid or certain
alcohols and oils, with a small amount of paraffin
or other hydrocarbon added in order to stabilise
the froth.

If a mixture of ore and quartz is added to a
liquid of this kind and the whole mass agitated
until froth is produced, the mineral ore is held in
suspension by the froth while the quartz sinks to
the bottom. A little acid or soda is sometimes
added to assist in the separation, though with
some oils as aerators the addition of a stabiliser
is not needed.

The quantities of reagents required are small,

about 3 lb. of oil, up to 20 lb. of acid or 3 to 4 lb. of sodium silicate being used per ton of material treated. It is found that different materials require different solutions for their effective separation.

The crushed material is mixed with several times its weight of water and agitated, and the reagents are added. After the material has been in contact with the liquid for a sufficient length of time the froth is run off into settling tanks containing water, where the dirt falls to the bottom and the froth containing the valuable material is removed.

A similar process can be used for coal washing, in order to separate the shale and dirt, and a few plants are in operation. The following particulars were given by Messrs. Scoular and Dunglinson of a plant recently installed at Oughterside Colliery, which comprises 2 mixing boxes, 8 agitation cells and 8 froth boxes. The plant is built of wood with hard wood liners to the agitation boxes, though these are being replaced by cast iron as the wear and tear is considerable.

The washer is 37 ft. long by 10 ft. wide and 15 ft. high and has an average capacity of 15 tons per hour. Cresylic acid is used as the aerator, with petroleum gas oil as a froth drier to give stability to the froth.

The cost per ton of material treated worked out at 5·29 pence for materials and labour, while the power consumption was about 94 horse-power hours per ton, the plant using about 200 gallons of water per minute.

4*

With coal having an analysis of fixed carbon
38·3 per cent., volatiles 18·4 per cent. and ash
43·3 per cent., the washer gave the following
results :—70·4 per cent. of coal having an average
ash content of 5·25 per cent., 7 per cent. of coal
containing 14·66 per cent. ash and 22·6 per cent.
of tailings containing 75·5 per cent. ash.

The Froth Flotation process is particularly suitable for separating low grade materials which
cannot be concentrated economically by other
processes.   It is necessary to find out by experi-
ment the most suitable reagents and their correct
proportions which will ensure the more valuable
material being kept in suspension by the froth
while the material to be discarded is thrown down
in the settling tanks.

In some cases, as in the Elmore process, the
action of the froth is assisted by subjecting the
mixed pulp of ore and water, with the added oil
and acid, if used, to a partial vacuum, which
largely increases the volume of the bubbles.   The
particles of mineral which are coated with grease
by the selective action of the oil float on the
surface and can be removed easily.

The Elmore Vacuum Flotation plant is made in
this country by Messrs. Fraser and Chalmers,
Engineering Works, Ltd.

The capacity of a one-unit plant is about 40
tons per 24 hours, and the power required for its
operation is about 5 B.H.P.

CHAPTER VI

THE DRYING OF MATERIALS

Iᴛ is usually necessary, after materials have been cleaned or graded by any of the processes described previously where water is employed as the separating medium, to remove the surplus water and to dry them more or less completely according to the purposes for which they are required.

Where the materials are unaffected by heat, as in the case of sands and similar materials, the final drying can be carried out by means of hot air or by distributing the materials over surfaces exposed on the underside to the heat from furnace gases, but where the materials would be damaged by excessive heat, as in the case of coal, other means must be employed.

Coal after washing may be dried by simple drainage, but this process is slow and requires considerable space, and mechanical means greatly reduce the time and space needed for removing the surplus moisture.

Various types of Centrifugal Drying Machines are in use for removing the moisture from many classes of materials. In the simplest form a Centrifugal Separator or Centrifuge consists of a cylinder or basket supported by a central spindle and rotated rapidly by means of a belt drive or by an electric motor or steam turbine. The wet material is fed into the basket before starting up the machine in the case of very fine materials which require an inner lining to the basket of canvas or felt, or while the machine is running when a plain perforated cage is used.

107

The water is driven off by centrifugal force through the perforated walls, the holes in which must be sufficiently fine to retain the material. The efficiency of the drying depends on the speed of the basket and the time during which it is allowed to operate, the higher the speed and the longer the time the more complete being the separation of the water. The size of the material also affects the degree of separation. With fine material the speed should not be so great as is permissible with large particles, otherwise there is a danger of the particles being so tightly packed together by the centrifugal action that the moisture cannot penetrate the mass.

The amount of moisture extracted is not directly proportional to the centrifugal force. According to some figures recently published,* in one test on coal containing 21 per cent. moisture an increase of speed giving three times the centrifugal force only increased the extraction by 38 per cent., while in another case where the coal contained 15 per cent. moisture the amount extracted only increased by $14\frac{1}{2}$ per cent.

There are various types of Centrifugal Separators in commercial use, some depending solely on centrifugal action and others arranged so that in addition a stream of air is drawn through the material. The moisture is reduced generally to about 7 to 8 per cent., according to the size of the material, as the larger the general size of the pieces the smaller will be the percentage of moisture remaining, and the ordinary type of Centrifugal

* *Colliery Engineer*, April 1924.

Separator is therefore most useful for preliminary drying to a point which is sufficient for many purposes. In those cases where complete extraction of the moisture is necessary some other form of dryer must be used.

Centrifugal machines are made in capacities up to 85 or 90 tons per hour, and the power required varies from about $\frac{1}{2}$ to $\frac{3}{4}$ B.H.P. per ton of material handled per hour. More power is required when starting up than when running at full speed.

The speed is regulated so that the centrifugal force on the material is about 100 to 130 lb.

When dealing with fine material a small portion, approximately 2 per cent., will pass through the screen with the water.

A modified type of Centrifuge may be used for grading materials in a wet state. A deeper basket is used with circular zones of different-sized holes, the larger openings being at the top. The centrifugal action drives out the material with the water into annular chambers surrounding the basket, from which the various sizes of materials can be led through shoots to separate bins.

For the preliminary drying of sands and similar materials settling tanks are frequently used, the sand and water being run into large tanks or ponds and, after a sufficient time has elapsed for the material to settle, the water is run off through sluices and the material dug out or removed by bucket elevators, or conveyors of the scraper or worm types. The buckets are usually perforated and thus act as secondary drainers, while the conveyors also allow a certain amount of water to drain back into the tank.

The labour required for digging out the material and loading it into trucks or into the elevator is considerable, and where sedimentation is employed it is often more convenient to use large inverted cones provided with a valve at the apex through which the sand can be discharged.

In some patterns of settling tanks the opening of the outlet valve is entirely automatic and is governed by the weight of material run into the tank or the weight of the sand which has settled out. As soon as a sufficient amount of sand has run off the tank rises and the valve is closed.

The material discharged from settling tanks of the types described above is not dry, but is only freed from surplus water, and in many cases further drying is required before the material is ready for use.

Filter presses are sometimes used for removing further moisture from materials which have been partially dried by settlement. The damp material is run or forced into bags of stout canvas which are placed in a frame and subjected to heavy pressure by screw gear, hydraulic rams, or other mechanical means. This process leaves the material in solid cakes and is not in general use, except for some special materials such as gold slimes.

Metalliferous sands and slimes are frequently dried by means of Vacuum Filters.

As mentioned above, wet materials can be dried on open floors under which a series of flues are constructed through which hot gases from a boiler or furnace are drawn by means of a chimney or

mechanical draught.   This method is intermittent
and requires a considerable amount of labour for
distributing and removing the material, and
although simple is therefore principally suitable
for comparatively small plants.

A simple form of Mechanical Dryer in frequent
use consists of a rectangular wooden tower up
which a blast of hot air is driven by a fan.   The
material to be dried is carried to the top of the
tower by an elevator and falls down through the
stream of hot air.   In order to ensure thorough
contact between the material and the hot air
the sides of the tower are fitted with sloping boards
placed alternately on either side, the material
falling from one to the other in its passage down
the tower.   The dry material is taken away from the
foot of the tower by a conveyor or second elevator.

This form of Dryer is used satisfactorily for the
final drying of sulphate of ammonia and similar
materials, as well as for sands, after the bulk of
the moisture has been removed in a Centrifuge.

In a somewhat similar apparatus, made by
Messrs. Hardy and Padmore, Ltd., heated trays
are used, the material being scraped from one to
the other through openings by means of blades
carried on a revolving shaft.   A series of heated
pipes may be used in place of the sloping boards
or trays, and in this case no scrapers are required.

Rotary Dryers

Rotary Dryers are very efficient and are suitable
for handling large quantities of material.   They

are in extensive use in many industrial processes
and are continuous in action, as the wet material
is fed in from the storage bin at one end by means
of a shoot or elevator and is delivered quite dry
at the other end.

This form of Dryer usually consists of an inclined
or horizontal cylinder which is rotated on its axis
in a similar manner to a rotary screen, by means
of a central shaft or external gears, and is heated
either by an external furnace in which it is supported or by means of a stream of hot air drawn
through it by a centrifugal fan.

Horizontal cylinders may be used, and in that
case the cylinder is sometimes fixed and the
motion of the material effected by means of a
revolving worm or by inclined blades or paddles
fixed to, and actuated by, a central shaft. Pro-
jecting angles are usually fitted to the inside of
the revolving cylinder to break up the mass of
material and to ensure thorough contact of the
particles with the heated air, or with the hot
walls of the cylinder in the case of externally-
heated dryers.

The Ruggles-Coles Rotary Dryer, made by the
Boving Engineering Works, Ltd., consists of a
double inclined cylinder, separated by suitable
spiders, the outer cylinder being supported on
rollers and rotated by means of annular gearing.
The wet material is fed into the annular space
between the two cylinders, projections being fitted
to ensure the thorough breaking up of the mass
and repeated contact of the material with the two

cylinders. Heated air is drawn through the drum by means of a fan fixed near the upper end, this air passing into the centre tube at the higher end, through this cylinder to the lower end, where it returns through the annular space to a fixed section at the upper end to which the fan inlet is connected. Dampers are provided for regulating the temperature by the admission of more or less cold air according to the results desired. A certain amount of very fine material is drawn through the cylinders with the gases, and the fan should be of heavy construction, to withstand the cutting action, if the material is at all abrasive in its nature.

The Sturtevant Rotary Dryer is of the single shell type, heated internally and externally by boiler flue gases or gases from coal-, coke-, gas- or oil-fired furnaces where waste gases from boilers are not available, diluted with cold air if necessary to produce the most suitable drying temperature. The cylinder is arranged horizontally in order to avoid axial thrust, and the material is fed through the Dryer by means of inclined blades or lifters.

Fig. 22 shows the arrangement of a direct-fired Dryer. The wet material is fed in at one end, the inlet for the heated gases being arranged at the same end, together with a cold air inlet fitted with a damper for regulating the temperature. The gases are drawn through the Dryer by means of a fan arranged on top of a casing built over the discharge end, the outlet from the fan being taken back to a brick chamber surrounding the body of the Dryer. This chamber serves to prevent loss of heat by radiation from the shell and also acts as a settling chamber for collecting any fine dust drawn through the apparatus with the gases.

If necessary a cyclone can be interposed between the fan and the air-jacket chamber in order to collect the dust if this is of any value or is present in such quantities that its removal from the air-jacket chamber would entail excessive expense.

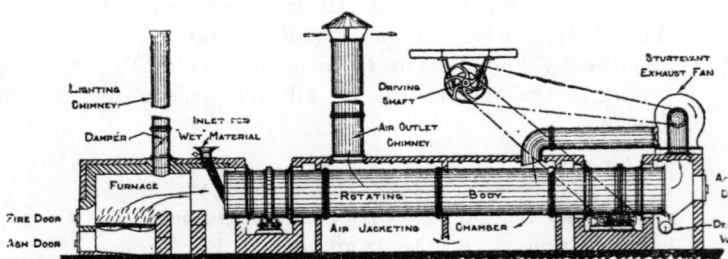

Fig. 22.—Sturtevant Rotary Dryer, using Direct-Fired Furnace and Air-Jacketing Chamber.

Where waste gases are available in sufficient quantity the air jacket may be dispensed with, and if the material is of such a nature that it would be injuriously affected by furnace gases, steam-heated coils may be used for raising the air to the desired temperature.

The cylinder is rotated by an annular toothed ring bolted to the casing, and driven through spur and bevel gears. It is supported on rollers at either end by means of circular treads bolted to the exterior of the shell, one of these treads having a flanged rim in order to locate the cylinder and

the other having a flat surface in order to allow
for expansion and contraction with variations of
temperature. The bearings of the rollers are
carried on inclined plates, and held in position by
screws, to allow of accurate adjustment of the
axis in a horizontal position.

Rotary Dryers of the types described above are
suitable for dealing with a very wide range of
materials, providing they are not of such a nature
that they will be damaged by the repeated lifting
and falling to which they are subjected.

Similar Dryers are made by several other firms
embodying modifications in the direction and travel
of the heated air.

A form of Dryer in use for the preparation of
pulverised coal consists of a vertical chamber
containing a number of cells through which the
coal passes. These cells are made slightly larger
at the bottom than at the top to prevent any
hanging up of the coal and are constructed with
perforated walls. The cells are arranged round a
central duct into which hot gases are led from a
suitable furnace or from the boiler plant, while
exhaust connections are arranged in the outer
casing. The coal is extracted from the cells by
means of revolving bladed drums fixed at the
lower ends. As dry coal is drawn off by the
extractors fresh moist coal is fed in by gravity
from hoppers at the top.

The hot gases are drawn through the apparatus
by means of an induced draught fan and pass first
through the inner perforated walls of the cells,
then through the descending mass of coal and

through the outer walls of the cells into the exhaust ducts. The temperature is regulated and over-heating of the coal prevented by the admission of cold air, and the moisture in the coal can be reduced in this type of Dryer to 2 to 3 per cent.

In another pattern the coal is distributed into the cells by means of a feeder, which regulates the supply evenly into the various compartments.

This type of Dryer has the advantage of occupy-ing little ground space, and the large area exposed to the flow of the gases reduces the velocity sufficiently to prevent any dust being carried over with the gases.

CHAPTER VII

THE dry cleaning of coal and the separation and grading of fine materials by means of air is now receiving considerable attention. The various processes in use all depend on the fact that particles of material of a given size and specific gravity are held in suspension by a current of air of a definite velocity, and by varying the velocity of the air the various grades of material can be separated and deposited where required.

The air velocity necessary to keep the particles in suspension in a vertical current is given approximately by the formula :—

$$V = K\sqrt{D\bar{S}}$$

where V = velocity of air in feet per second,
D = diameter of particles in inches,
S = specific gravity of the material,
K = a constant which varies with the shape of the particles and for approximately round grains may be taken as 77.

In metric units :

$$Vm = Km\sqrt{DmS}$$

where Vm = velocity in metres per sec.,
Dm = diam. in millimetres,
Km = 4·66 for rounded particles.

Ordinary expansion chambers are often used for the separation of metallic oxide dusts and fine cement from the furnace or kiln gases. These

117

chambers must be of ample size in order to reduce
the velocity of the air to a point where the finest
particles will be deposited, and are frequently
divided into a number of compartments with
sloping bottoms and doors so arranged that the
dust can be withdrawn without interrupting the
operation of the plant.

It may be noted that repeated changes of
velocity alone will separate out most materials
that can be conveyed by pneumatic means, but
the space occupied by the settling chambers will
in most cases be excessive, and the cyclone forms
a much more convenient arrangement.

Cyclones form an effective and simple means
of collecting dust and fine materials of various·
kinds which can be handled satisfactorily by
pneumatic means, though in their ordinary form
they are not able to separate out the particles
into different grades.

The cyclone consists essentially of a circular
chamber, usually constructed of steel plate, though
in large sizes they are sometimes built in reinforced
concrete, the upper portion of the chamber having
parallel or outwardly tapering sides, while the
lower portion is conical, tapering down to an outlet
pipe which is closed normally by a valve or dust
gate.

The dust-laden air enters through an inlet
connection which is arranged tangentially at the
upper edge of the cylindrical portion, and leaves
through a central outlet pipe which passes through
the top plate of the cyclone. This outlet is usually
protected by a cowl or hood, sometimes made

adjustable, in order to keep out rain, and may be of any convenient length.

For some duties the interior of the cyclone is fitted with a spiral of steel angle or plate, and where the material is abrasive it is advisable to protect the body of the cyclone with renewable plates of a hard material, otherwise the life of the cyclone will be very short.

The efficiency of a cyclone depends partly on centrifugal action and partly on sudden change of velocity. The air on entering is constrained to follow a circular path until it reaches the level of the lower end of the outlet pipe, which projects for some distance into the cyclone, where it flows inwards and out through the central pipe. As the air enters the body of the cyclone, which has an area much greater than that of the inlet pipe, the velocity is reduced until it is below that necessary to support the material, which is being thrown outwards at the same time by the centrifugal force, and it consequently falls down the sides of the conical portion, from which it is removed through the discharge pipe.

The diagram and table below give general dimensions which have been found effective for a variety of purposes, but it is necessary to use the dimensions with discretion, as in order to be fully satisfactory the air speed must be adjusted to suit the material which is being dealt with. The usual velocity through the inlet of the cyclone is from 30 to 50 ft. per second, according to the specific gravity of the material :—

Size of Fan Inlet.	A.	B.	C.	D.
9	36	57	6	39
12	48	76	8	52
15	60	94	8	64
18	72	113	9	77
21	84	130	$10\frac{1}{2}$	89
24	96	150	12	102
27	108	169	$13\frac{1}{2}$	115
30	120	188	15	128
33	132	206	$16\frac{1}{2}$	140
36	144	227	18	153

All dimensions in inches.

In this connection the following approximate air speeds used for conveying various materials may be of interest :—

Wood refuse .	velocity	= 3000 ft. per minute.		
Sawdust . .	,,	= 1200	,,	,,
Metal dust .	,,	= 1800	,,	,,
Fine brass turnings	,,	= 4000	,,	,,
Fibrous jute dust	,,	= 2000	,,	,,
Rubber dust .	,,	= 2000	,,	,,
Grain . . .	,,	= 3000	,,	,,
Fine coal . .	,,	= 4000	,,	,,

Where the collected material is of a combustible nature it can be delivered to the furnaces of any convenient boilers, or it can be run off into trucks or sacks as required.

In American cement manufacturing works, where waste heat boilers are being largely installed to recover the heat in the waste gases from the kilns which would otherwise be lost through the chimney stacks, cyclones are being fitted to collect the fine cement dust which is drawn through the boilers and economisers. This dust has an appreciable value as it consists largely of the finest particles

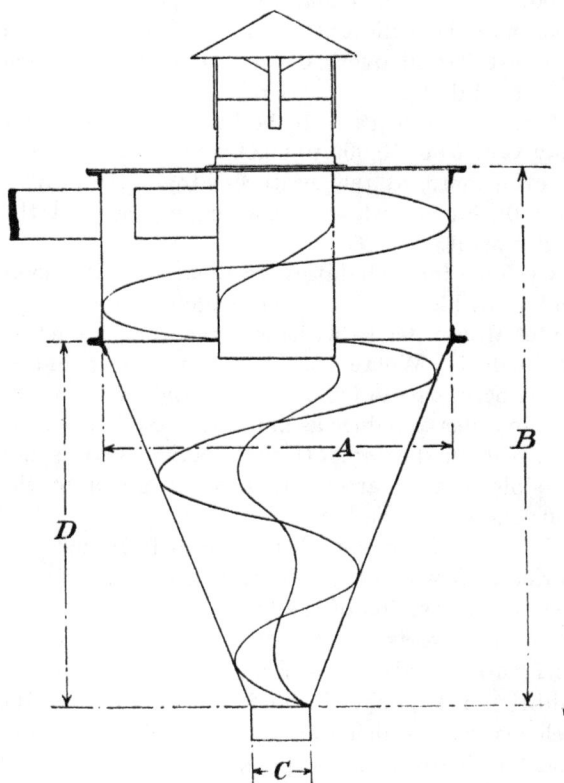

Fig. 23.—Cyclone.

of cement, though it may be contaminated to some extent by ash from the coal used for firing the kilns. Apart from other considerations, it is therefore quite worth while to save the dust, which otherwise becomes a nuisance to the surrounding neighbourhood. It is claimed that 90 to 97 per cent. of the dust may be collected in an efficient plant. In one installation over 80 tons of dust have been recovered daily.

The volume of gases to be handled in each unit may vary from 40,000 to 60,000 cu. ft. per minute, or even more, at temperatures ranging from 350° to 450° F., according to the arrangement of the boiler plant.

Cyclones for such duties are about 20 ft. diameter and 40 ft. high, and must be made of heavy gauge material in order to withstand the abrasive action of the dust. Where possible the cyclone should be fitted between the boiler and the induced draught fan, in order to reduce as much as possible the wear and tear on the fan. If this arrangement is not possible, special attention must be given to the construction of the fan.

Cyclones have also been successfully used in various gas works for separating out from the gas the heavier portion of the tar which is carried over from the retorts. The usual pattern of cyclone, as illustrated above, is used, sometimes with the addition of a damper in the inlet for regulating the velocity with which the gas enters, but the lower end terminates in a seal pipe, to prevent escape of gas, instead of the usual dust gate or valve. It is necessary for the gas to be cooled down to a temperature of about 100° to 120° in order that the tar particles may be condensed sufficiently for the centrifugal action to be effective.

A number of types of dry cleaning plants using air as the separating medium have been designed for the cleaning of coal and the grading of fine materials which are too small to be dealt with by the ordinary type of screen.

If a fine material is carried in a current of air and the speed of the air reduced by a certain amount, the finer particles will be carried forward and the coarser or heavier particles will be deposited, as explained above.

The various principles in these machines may be summarised as follows :—

(1) Reciprocating or vibrating devices with continuous air currents.

(2) Reciprocating or vibrating devices with pulsating air currents.

(3) Stationary devices with continuous air currents.

(4) Stationary devices with pulsating air currents.

The first type mentioned above is the most successful, and most of the modern Air Separators operate on these lines. Various plants are being installed in America for dry cleaning coal having capacities up to 150 tons per hour.

The presence of clay or similar plastic matter prevents the effective use of Air Separators, but for fine granular material this system is very efficient.

A current of air having a velocity of about 4·5 ft. per second will hold in suspension and carry away the particles less than 0·1 millimetre diameter.

Air Separators of type 2 above are used for concentrating metalliferous sands. A vibrating table or frame covered with fabric is used, the sand being fed on to the table while a pulsating current of air is introduced underneath.

The air current causes the material to arrange itself in layers, the lightest uppermost, while the vibrating or reciprocating action causes the various grades of sand to arrange themselves in bands or zones. The Sutton-Steele Separator is of this type.

Several Air Separators have been designed similar in shape to a cyclone, but with a double casing generally constructed of sheet steel, in the centre of which is arranged a revolving spreader and a fan wheel. The raw material is fed into the centre of the apparatus, where it falls on to the spreader and is thrown outwards in a thin sheet. Air from the fan is blown or drawn through this sheet of material and carries with it into the outer casing the fine material, while the coarser particles fall into the outer casing from which they are discharged separately.

As the size of the particles carried forward by the air depends solely on the velocity of the air stream, the fineness of the separated material can be regulated by throttling the air supply to the fan wheel, or by regulating the speed of the fan.

This type of Separator was originally introduced by Mumford and Moodie, but is now made by other firms, and various modifications have been made in the details in order to increase the efficiency of separation. The improvements consist mainly in dividing up the air stream by baffles or other means in order to increase the separating effect.

Air Separators are most effective with very fine

material and will not handle satisfactorily particles over $\frac{1}{8}$ in. in diameter. The material must be dry so that the particles will not ball, or hang together, and the amount of material fed in should be regulated so as not to give too thick a stream for the air to pass through.

The air leaving the apparatus may contain a quantity of very fine dust which should be collected in cyclones or settling chambers. If the dust is valuable bag filters may be used in addition, or otherwise a tower scrubber may be installed.

The specific gravity as well as the size of the particles affects the separation, hence Air Separators of this type should not be used for material containing particles of widely differing specific gravity and size. If the material is approximately equal in size it will be graded according to the weight of the particles, or if all the material is of the same nature, the grading will be in accordance with the size of the various particles.

The Sturtevant Air Separator is a typical example of this type of apparatus and consists of an outer casing of steel plate, the upper portion being cylindrical and the lower portion forming a conical hopper.   Inside this casing and fixed rigidly to it are a cylindrical sleeve and a smaller conical hopper provided with an outlet spout passing through the side of the outer shell.   There is a central shaft, driven through bevel gearing on which are fixed a fan wheel, deflecting rings and a serrated cone. The material, from which large pieces should first be removed by screening, is fed through an opening in the top plate and falls on to the revolving distributing plate, from which it is thrown outwards by centrifugal force. A small deflector plate prevents large lumps passing into the moving parts.

The deflecting rings and serrated cone break up the material in order to allow the air to act freely on the particles. A stream of air is drawn by the fan through the falling material, carrying with it the fine particles which are whirled round in the outer casing. This casing forms a Cyclone Separator and allows the dust to settle out and collect in the conical portion, from which it is discharged through a central outlet. The larger particles are not carried forward by the air stream and fall into the inner hopper, from which they are discharged through a separate outlet.

The following table gives the capacity of various sizes of Sturtevant Air Separators, the fine product giving 90 per cent. through a 100-mesh screen :—

Diameter.	Approximate capacity.	Approximate Horse-power.	Pulley speed.
5 ft.	1 to 1½ tons per hr.	2 to 5	100 to 150
6 ft.	1½ to 2 ,, ,,	3 to 5	100 to 150
8 ft.	4 to 6 ,, ,,	4 to 6	250 to 300
10 ft.	6 to 8 ,, ,,	5 to 8	250 to 300
12 ft.	8 to 12 ,, ,,	8 to 10	350 to 400

The degree of fineness is regulated by a valve which controls the speed of the air through the material.

In the Selektor efficient separation is effected by arranging a number of annular plates below the spreader, the air carrying with it the finer particles being drawn between these plates into the inner casing, while the larger particles fall into the outer cone.

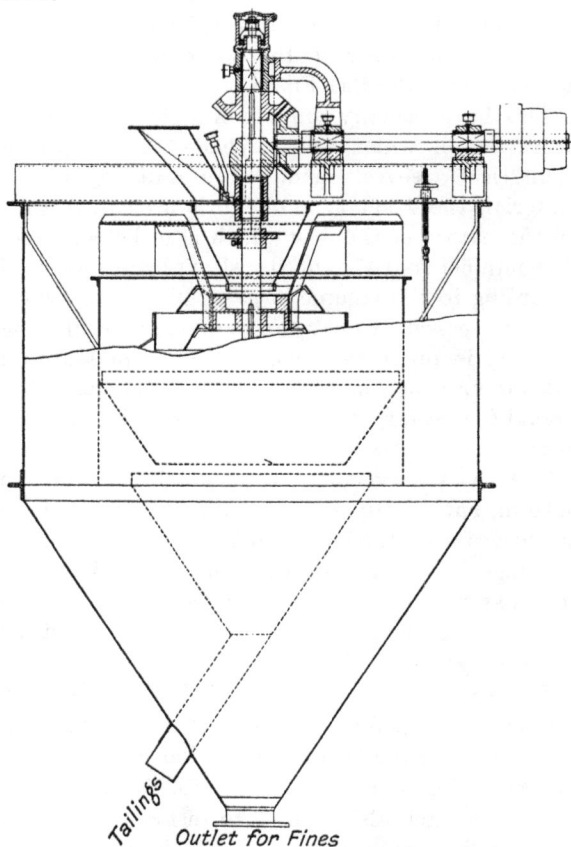

FIG. 24.—Sturtevant Air Separator.

The arrangement of a typical Air Separator is shown in Fig. 24.

Bag Filters

A type of Filter or Separator which is in extensive use in cement mills and other works where fine dust is produced is the Bag Filter.

This Filter usually consists of a steel framework or casing in which are suspended a number of cylindrical sleeves, made of a suitable textile material, the exact type of fabric used depending on the nature of the dust which is to be collected. The number and size of the sleeves are regulated according to the volume of gases to be dealt with, as it is necessary to have an ample area of filtering material in order to reduce the back pressure on the fan or other means used for producing the draught necessary to force the gases through the filter.

The gases may enter either at the top or the bottom, but usually at the bottom, where the bags or sleeves are attached to inlet rings surrounding openings in the top of the duct. The bags are closed at the top and are suspended from the upper part of the casing, where there is also an outlet for the clean gas.

The dust-laden air or gas enters the sleeves through the open ends and passes through the fabric, leaving the dust on the inside surface of the sleeves. Means are provided for shaking the sleeves at intervals in order to dislodge the dust, which falls into a hopper at the bottom and is removed thence through doors or by means of a worm conveyor.

In some installations the direction of the air-

flow through the fabric can also be reversed, to
facilitate the removal of the dust, and the filters
are arranged in sections so that one unit at a time
can be cleaned without interfering with the continuous operation of the plant.

Bag Filters are not suitable for hot or moist
gases, as the fabrics generally used will not stand a
high temperature, and the presence of moisture
causes the dust to cake on the sleeves and prevent
the passage of the gas, but for removing fine dust
from dry and cool gases they are very effective.

The average resistance offered by a well-designed
plant is about 2 in. water gauge.

In determining the number and size of the bags
the reduction in effective area due to the accumulation of dust must be borne in mind, and the fact
that one or more bags will usually be out of action
for cleaning.

It is usual to make the area of the filtering cloth
not less than 450 times the area of the fan outlet,
but the actual proportion will depend on the
amount of dust present in the gases.

Fig. 25 shows the arrangement of a typical small
unit.

The Halberg Beth Dry Gas Cleaning Plant has
been designed specially for removing the dust from
blast furnace gas and employs bag filters of the
type described above, together with additional
apparatus for cooling the gas. The dust is not
only detrimental to the use of the gas for combustion
in engines or under boilers, but contains a considerable proportion of valuable potash, which can
readily be recovered from the dry dust. The crude

5

gas usually contains from 6 to 10 grammes per cubic metre, and this can be reduced to 0·002 gramme or even less.

The filter bags are made of a cotton fabric with a raised surface or nap on one side. The bags are

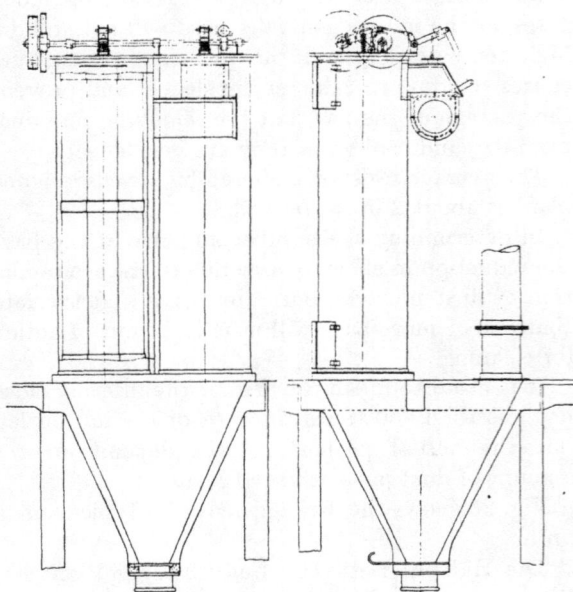

FIG. 25.—Sturtevant Bag Filter.

10 ft. long by 8 in. diameter, open at the bottom and supported at the top by a steel cap. Twelve bags are placed in a compartment constructed of steel plate and twenty of these compartments form a standard unit, which has a capacity of 1 million cubic feet of gas per hour. Smaller units can be

made up if necessary. The bags are shaken in a vertical direction every seven minutes, one compartment being cleaned at a time so that the operation of the filter is not interrupted, and the cleaning operation is facilitated by reversing the gas current while the bags are being shaken. The dust falls into a compartment arranged under the bags, from which it is removed by a worm conveyor into dust hoppers.

The gases are cooled down to about 100° C., which is the maximum temperature to which the filter bags can be subjected with safety. If the gases are liable to contain moisture they are slightly preheated before passing into the filter, in order to prevent the bags becoming damp. The delivery temperature is 70° to 80° C. when required for combustion in furnaces, but when used for gas engines the gas is further cooled to a temperature of 20° to 30° C. The power consumption is about 1 kilowatt for 25,000 cub. ft. of gas cleaned, together with a small amount of steam for the preheater, if waste heat from stoves or boilers is not available.

The Halberg Beth System is made in this country by Messrs. Fraser and Chalmers, Engineering Works, Ltd.

CHAPTER VIII

ELECTRICAL SEPARATION

THERE are two methods of separating materials by electrical means. One method is operated by magnetism produced generally by means of powerful electromagnets, while the other depends on the effect of electrostatic charges.

Whenever raw materials are liable to be contaminated by the presence of pieces of iron or other magnetic bodies, Magnetic Separators can be used. They have been found particularly useful for removing miners' picks, coupling links and pins, from coal passing to a crusher, in order to prevent damage to the machinery, and they are frequently used in stone quarries for a similar purpose. They are in considerable use in the manufacture of paper, cement, gypsum and limestone, and in the cleaning of grain and tobacco, where the presence of iron particles would reduce the value of the finished material. Magnetic Separators are also employed for separating iron from the combustible materials collected in municipal refuse plants before putting the bulk of the refuse into the destructors, and for removing iron from blast furnace slag, brass scrap, gold ore and many other raw materials.

Although iron is the best-known magnetic substance and is by far the most susceptible to the magnetic influence, many other substances, particularly iron ores, are more or less strongly attracted by a magnet and can be removed by means of a very strong magnetic field from other materials which are not affected by the magnetic influence.

For very small quantities of material a magnet

may be drawn through the mass and removed with the magnetic particles adhering to it, but this is not a commercial operation, as the capacity of such an arrangement is too limited.

In the simplest form of Magnetic Separator the raw material is fed on to a belt conveyor at the delivery end of which a special magnetic pulley is arranged. The magnetic field attracts the pieces

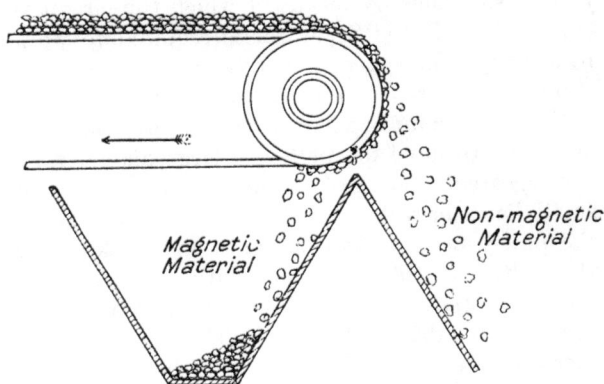

FIG. 26.—Magnetic Separator.

of iron and keeps them in contact with the belt as long as the belt remains on the surface of the pulley, while the non-magnetic material, which is not attracted, is delivered into a shoot, the pieces of iron falling into a separate shoot or bin as soon as they leave the influence of the magnetic field.

The action of a simple Magnetic Separator of this kind is shown in Fig. 26.

The magnetic field is set up by means of windings arranged in the interior of the pulley, through which a current of electricity is passed. Direct current is used, generally at voltages of 110 or 220, though higher voltages can be used if required. The pulleys consist of a number of steel discs mounted on a hollow shaft, on which is also carried a steel bobbin, secured firmly to the discs. The magnetising coils are wound on the bobbin and the connections are brought through the hollow shaft to a pair of slip rings by means of which the circuit is completed. If the Separator is required to work in a dusty atmosphere it is advisable to have the slip rings enclosed in a dustproof cover.

From the point of view of belt maintenance it is advisable to have the pulley of large diameter, as the repeated bending of the belt round a pulley of small diameter will soon ruin it. In no case should the pulley be less than 30 times the thickness of the belt, but this ratio can with advantage be increased, particularly if the magnetic pulley is also used for driving the belt. This should be the case whenever possible, as it is always advisable to drive a belt conveyor at the delivery end. If it is not possible to arrange the plant in this way a short separate conveyor should be installed as the separating device, on to which the main conveyor or elevator can discharge.

A large pulley has the additional advantage that for a given belt speed the angular velocity is reduced in proportion as the pulley diameter is increased, hence there is less tendency for the iron to be thrown off by centrifugal force against the attraction of the magnet, and the separation is carried out more perfectly. Against this increased

efficiency must be set the increased cost of the larger
pulley. It is important that the relation between
the size of the pulley and the windings should be
carefully proportioned in order to obtain the best
results for a given expenditure of power. The
current consumption is quite small.

The Igranic Electric Co., Ltd., give the following
figures for the size, current consumption and
capacity for some of their standard sizes of Magnetic
Separator pulleys :—

Dia. of pulley. Inches.	Width of belt. Inches.	Rating. Cubic ft. per hour.	Watts.
8	12	670	187
8	18	1000	300
8	24	1330	413
12	18	1500	448
12	24	2000	552
12	36	3000	942
15	24	2000	626
15	36	3000	1370
15	48	4000	1625
18	24	2000	940
18	36	3000	1440
18	48	4000	1960
24	24	2000	1230
24	36	3000	1910
24	48	4000	2600
24	60	5000	3280
30	24	2000	1565
30	36	3000	2420
30	48	4000	3260
30	60	5000	4100

The capacities listed above are based on the usual
depth of material of about 3 in., with the conveyor
belt running at a speed of 100 ft. per minute.
Some materials may be handled with a higher belt
speed and a greater depth on the belt, but other
materials can only be handled efficiently at a

reduced rate, and the most suitable belt speed and depth of feed can only be determined by experience.

If the separating drum revolves between the poles of a fixed magnet, the strength of the magnetic field can be made variable, with its strongest point exactly opposite the magnet poles. The strength of the field falls off as a point on the circumference of the drum recedes from the magnet. By means of this arrangement the particles which are most strongly attracted by the magnetic field can be carried further round the drum and separated from the less magnetic material, separate hoppers or shoots being provided for the different grades.

The details given above relate to the simplest types of Magnetic Separators. For more intensive operations, such as the concentration of ores and for separating roasted sulphide ores, blast furnace dust and materials of a similar nature, a somewhat different system is used, employing stronger magnetic fields.

In one type of Separator the material is carried by a band conveyor underneath one or more revolving iron discs, which are magnetised locally by suitable coils, the lines of force being concentrated along the line of the belt by a steel bar fixed below the belt. The revolving discs have their edges notched, leaving projections which form auxiliary pole pieces. As the disc revolves the magnetic material is attracted by the strongly magnetised poles of the disc, but as these poles pass outside the magnetic field they lose their magnetism and release the material on to shoots arranged on either side of the machine.

A second disc having knife edges formed on its lower side is arranged in a further magnetic field of higher intensity, in order to remove the less magnetic material.

Fig. 27 shows diagrammatically the arrangement of one of these Separators.

A similar type of Separator, with one disc and the material conveyed by means of a vibrating conveyor instead of a belt, forms a most efficient apparatus for removing iron particles from brass

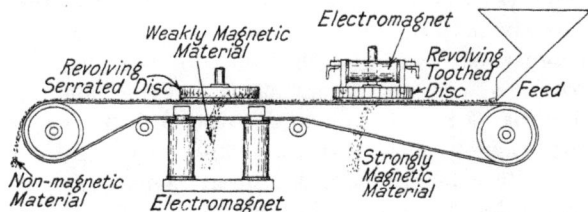

FIG. 27.—High Intensity Magnetic Separator.

and similar scrap material. The feed can be regulated to any desired amount and the material passes through two strong magnetic fields before it leaves the conveyor.

The double magnet machine is particularly suitable for ore separation and concentration and is used for dealing with roasted sulphide ores, zinc blende, copper, nickel and iron ores and blast furnace dust.

Three high intensity magnets may be used in series where difficulty is experienced in concentrating the ores or where more than one magnetic constituent is to be extracted.

In the Wetherall Magnetic Separator the material is carried by a short belt conveyor and is carried between the poles of three electro-magnets, round one of which the belt passes. The magnetic material is carried round the one magnet where it is deflected by the magnetic field and falls into a hopper, while the non-magnetic portion is discharged between the other two magnets into a separate hopper.

The Monarch Separator is of the drum type, but employs two magnetic drums of different intensities. The raw material falls first on to the more strongly magnetised drum and the whole of the magnetic particles are extracted. This portion of the material is then fed on to the second drum, where the magnetic field is weaker, and only the more strongly magnetic particles are attracted. In this way a double classification can be effected.

According to A. B. Searle, a machine of this type with drums 2 ft. diameter rotating at 40 and 50 r.p.m. will handle 15 to 20 tons of material per hour with an expenditure of 2 to 3 H.P. for the electromagnets and $\frac{1}{2}$ to $\frac{3}{4}$ H.P. for revolving the drums.

Electrostatic Precipitation or Separation, which is the method employed in the second class of Electrical Separators, depends on the fact that when materials which are conductors of electricity are brought under the influence of a strong electric field, they become charged with electricity similar to that in the primary conductor which is setting up the electric field and are consequently repelled by that conductor.

Electrical Precipitators of this type have been applied to the purpose of separating valuable metallic dust from furnace gases and are in considerable use for cleaning blast furnace gas to enable the gas to be used for internal combustion engines or for burning under steam boilers. They have also been used for the removal of tar fog from ordinary towns' gas.

Blast furnace gas contains from 5 to 15 grammes of dust per cubic metre, and for satisfactory combustion in gas engines the amount of dust should be reduced to about 0·02 gramme per cubic metre, though for burning under boilers a larger amount may be permitted.

The usual type of Electrostatic Separator consists of a large number of vertical pipes, from 6 in. to 9 in. diameter and about 10 ft. long, with a chain or wire running down the centre of each pipe, this chain or wire being charged electrically so that an electrostatic field is set up in each pipe.

The gases are admitted at the lower ends of the pipes and as they pass upwards the electrical action causes the dust to settle on the wires and tubes. At intervals of half to one hour, according to the quantity of dust in the gas, the flow of gas is shut off and the tubes and wires are shaken or rapped, causing the dust to fall to the bottom of the Separator chamber. This operation takes from one to two minutes. Usually a number of units are arranged in each plant so that one unit at a time may be shut down for cleaning without interfering with the supply of gas.

This method of gas cleaning is economical as

about 3 kilowatts only are required for cleaning 1 million cubic feet of gas.

Electrical Precipitators have been applied successfully to the recovery of dust from copper, lead, tin and aluminium furnaces, as well as to the removal of dust from blast furnace gas and the gases from coal-briquetting plants.

High tension current is used, at voltages up to 100,000, the electrical plant consisting of a motor generator and transformer, rectifier and control panel.

In one copper smelting works, where the electrical precipitation plant consists of 640 pipes, 6 to 10 tons of dust are recovered per day, this dust containing an appreciable amount of copper and large quantities of arsenic.

In another type of Electrostatic Separator the material is fed on to a positively-charged drum by means of a shoot which is charged negatively, and as the drum revolves the material passes between the drum and a second negative pole.

The particles which are good conductors are immediately charged positively and are repelled towards the negative pole, while the inferior conductors, being less strongly affected, adhere to the drum until they are beyond the influence of the negative pole, while the material which is non-conducting adheres until it is scraped off in some convenient manner. It is thus possible to separate the material into three or more grades, according to their relative conductivities.

Thanks are due to those firms who have kindly furnished details of their special types of machinery.

There are many other machines which might have been described, but the author hopes that the general survey of the subject given in the foregoing pages will be of use to those interested in the Screening and Grading of Materials.

INDEX

www.ingramcontent.com/pod-product-compliance
Lightning Source LLC
Chambersburg PA
CBHW021425180326
41458CB00001B/133